6

NUCLEAR MATERIALS AND DISASTER RESEARCH

NUCLEAR ENERGY, FACILITY SITING AND WASTE STORAGE

PUBLIC ATTITUDES AND PREFERENCES

NUCLEAR MATERIALS AND DISASTER RESEARCH

Additional books in this series can be found on Nova's website under the Series tab.

Additional E-books in this series can be found on Nova's website under the E-book tab.

NUCLEAR MATERIALS AND DISASTER RESEARCH

NUCLEAR ENERGY, FACILITY SITING AND WASTE STORAGE

PUBLIC ATTITUDES AND PREFERENCES

FRANCIS CROWSON
EDITOR

nova publishers
New York

For permission to use material from this book please contact us:
Telephone 631-231-7269; Fax 631-231-8175
Web Site: http://www.novapublishers.com

Library of Congress Cataloging-in-Publication Data

ISBN: 978-1-62948-787-8

Published by Nova Science Publishers, Inc. † New York

CONTENTS

PREFACE

The United States (US) program for siting interim storage and permanent disposal facilities for used nuclear fuel (UNF) is at a crossroads. The March 2010 request by the US Department of Energy (DOE) to the US Nuclear Regulatory Commission (NRC) for termination of the Yucca Mountain Project (YMP) license application, followed one year later by the disastrous nuclear events in Fukushima, Japan, have resulted in a fundamental reconsideration of approaches for siting interim and permanent UNF management facilities in the US. This book provides findings from a set of social science studies undertaken by the Center for Risk and Crisis Management (CRCM) and Sandia National Laboratories (SNL), which focus on public attitudes and preferences concerning the siting of nuclear repositories and interim storage facilities. This book is also a framework for moving toward a sustainable program to deploy an integrated system capable of transporting, storing, and disposing of used nuclear fuel and high-level radioactive waste from civilian nuclear power generation, defense, national security and other activities.

Chapter 1 – This report provides findings from a set of social science studies undertaken by the Center for Risk and Crisis Management (CRCM) and Sandia National Laboratories (SNL), which focus on public attitudes and preferences concerning the siting of nuclear repositories and interim storage facilities. Overall these studies are intended to be responsive to the recommendation of the Blue Ribbon Commission on America's Nuclear Future (BRC) that US Department of Energy (DOE) learn as much as possible from prior experience. As stated by the BRC (BRC 2012: 118): To ensure that future siting efforts are informed by past experience, DOE should build a data base of the experience that has been gained and relevant documentation produced in efforts to site nuclear waste facilities, in the United States and abroad…

Specifically, this report describes the findings from four new studies undertaken in 2012, including (1) an Internet survey conducted in June 2012, with 2017 adult residents of the continental US focused on nuclear issues and nuclear facility siting; (2) an analysis of the outcomes of 269 cases of attempted nuclear facility siting efforts globally spanning 31 countries over 50 years; (3) trend analysis of evolving nuclear sentiment in the US, employing a total 287 questions drawn from dozens of nationwide surveys from 1973 through 2011; and (4) a time-series study, utilizing the content of social media and patterns of online information searches in 2010-2011, to analyze the changes in public attention to nuclear energy and nuclear waste that followed the Fukushima nuclear event in March of 2011. These studies add to the stock of knowledge that will facilitate the transition to a consent-based siting program for interim storage and permanent disposal facilities for used nuclear fuel (UNF) and high-level waste (HLW) in the US.

Chapter 2 – The *Strategy for the Management and Disposal of Used Nuclear Fuel and High-Level Radioactive Waste* is a framework for moving toward a sustainable program to deploy an integrated system capable of transporting, storing, and disposing of used nuclear fuel and high-level radioactive waste from civilian nuclear power generation, defense, national security and other activities.

The Strategy addresses several important needs. First, it serves as a statement of Administration policy regarding the importance of addressing the disposition of used nuclear fuel and high-level radioactive waste; it lays out the overall design of a system to address that issue; and it outlines the reforms needed to implement such a system. Second, it presents the Administration's response to the final report and recommendations made by the *Blue Ribbon Commission on America's Nuclear Future* ("BRC"). It also responds to direction in the Joint Explanatory Statement accompanying the Consolidated Appropriations Act, 2012, to develop a strategy for the management of used nuclear fuel and nuclear waste in response to the BRC's recommendations. Third, this strategy represents an initial basis for discussions among the Administration, Congress and other stakeholders on a sustainable path forward for disposal of nuclear waste.

In: Nuclear Energy, Facility Siting … ISBN: 978-1-62948-787-8
Editor: Francis Crowson

Chapter 1

PUBLIC PREFERENCES RELATED TO CONSENT-BASED SITING OF RADIOACTIVE WASTE MANAGEMENT FACILITIES FOR STORAGE AND DISPOSAL: ANALYZING VARIATIONS OVER TIME, EVENTS, AND PROGRAM DESIGNS*

Hank C. Jenkins-Smith, Carol L. Silva, Kerry G. Herron, Kuhika G. Ripberger, Matthew Nowlin, Joseph Ripberger, Evaristo "Tito" Bonano and Rob P. Rechard

EXECUTIVE SUMMARY

This report provides findings from a set of social science studies undertaken by the Center for Risk and Crisis Management (CRCM) and Sandia National Laboratories (SNL), which focus on public attitudes and preferences concerning the siting of nuclear repositories and interim storage facilities. Overall these studies are intended to be responsive to the

* This is an edited, reformatted and augmented version of the U.S. Department of Energy, Nuclear Fuel Storage and Transportation Planning Project, FCRD-NFST-2013-000076 SAND 2013-0032P, dated February 2013.

recommendation of the Blue Ribbon Commission on America's Nuclear Future (BRC) that US Department of Energy (DOE) learn as much as possible from prior experience. As stated by the BRC (BRC 2012: 118):

> To ensure that future siting efforts are informed by past experience, DOE should build a data base of the experience that has been gained and relevant documentation produced in efforts to site nuclear waste facilities, in the United States and abroad...

Specifically, this report describes the findings from four new studies undertaken in 2012, including (1) an Internet survey conducted in June 2012, with 2017 adult residents of the continental US focused on nuclear issues and nuclear facility siting; (2) an analysis of the outcomes of 269 cases of attempted nuclear facility siting efforts globally spanning 31 countries over 50 years; (3) trend analysis of evolving nuclear sentiment in the US, employing a total 287 questions drawn from dozens of nationwide surveys from 1973 through 2011; and (4) a time-series study, utilizing the content of social media and patterns of online information searches in 2010-2011, to analyze the changes in public attention to nuclear energy and nuclear waste that followed the Fukushima nuclear event in March of 2011. These studies add to the stock of knowledge that will facilitate the transition to a consent-based siting program for interim storage and permanent disposal facilities for used nuclear fuel (UNF) and high-level waste (HLW) in the US.

The key study findings from the nationwide Internet survey reported in Section 5 were the following:

- Perceived risks and benefits of nuclear energy are nearly equally balanced, and support for additional nuclear reactors is divided. Support for continued reliance on nuclear energy was suppressed by concerns raised by the Fukushima nuclear event. The most potent predictors of support for nuclear energy are perceived risks and benefits, with the risk of reactor accidents being the most prominent risk. Among the perceived benefits, the most important is reduced dependence on foreign energy sources. Greater trust in federal agencies (DOE, Environmental Protection Agency (EPA), and Nuclear Regulatory Commission (NRC)) to provide accurate information about nuclear risks also leads to greater public support for nuclear energy.

- Greater concerns about climate change, on average, lead to *less* support for nuclear energy. In part this results from the deeply held values that underlie nuclear preferences; more generalized concerns about the environment, and egalitarian worldviews, both contribute to beliefs that the climate is changing due to greenhouse gasses of human origin. Egalitarianism (indirectly) and concerns about nature (directly *and* indirectly) reduce support for nuclear energy.
- Public preferences for a national strategy for managing UNF favor pursuit of two permanent geologic repositories over continued on-site storage. Preference for interim storage falls between, with a plurality of respondents in support.
- Support for either a geologic repository or an interim storage facility is increased when the facility is co-located with a nuclear safety research laboratory, or would permit construction of a UNF reprocessing facility. More modest gains in support are evident when substantial financial incentives are offered to the prospective host state and community.
- A slight majority of respondents favored a "bottom-up" siting strategy wherein potential host communities nominate themselves for consideration over a "top-down" strategy in which experts identify technically optimal sites and then invite affected communities to consider hosting UNF storage and disposal facilities.
- Survey respondents indicated greatest trust for risk information provided by experts from the National Academy of Sciences (NAS) and national laboratories. Federal agencies (NRC, DOE, and EPA) also received relatively high marks on trust. At the same time, respondents viewed all organizations as prone to either downplay risks (industry groups, DOE, NRC, national labs) or exaggerate them (environmental advocacy groups, EPA) except for the NAS.
- When asked about the process by which consenting communities may consider hosting a UNF storage or disposal site, majorities of respondents believed that citizens (via referenda) and governors should be able to veto consent. Majorities of respondents opposed allowing other actors (federal elected officials, federal agencies, nongovernmental organizations, NGOs) to have a veto on consent.
- Respondents supported allowing potential host communities and states to withdraw from the siting process through the stage at which a license is submitted to federal agencies for review; majorities opposed permitting potential hosts to withdraw after a license is issued.

- When asked whether they would support siting a hypothetical interim storage facility or permanent repository, support was conditional on distance. Support was reduced the nearer the facility would be to the respondents' residence. However, when respondents were apprised of their current proximity to temporary UNF storage, those who currently live within 25 miles of a facility were likely to express greater support than those who lived farther from existing storage.

Among the key findings of our international study of past siting efforts reported in Section 2 are the following:

- All else being equal, the probability that a proposed nuclear facility will be completed and operational has decreased substantially over time, from near certainty in the mid-1950s to a fifty- fifty proposition for those siting efforts that had been concluded.
- Variation in the institutional frameworks for decision-making within countries explains a substantial fraction of the differences in siting outcomes: more democratic countries, and those with federal (decentralized) decision-making structures, have lower likelihoods for nuclear facility siting than countries that are less democratic and more centralized.
- Why do countries with greater democratic openness have a more difficult time siting nuclear facilities? The analysis indicates that greater democratic openness is associated an increased probability of expressed opposition to the facility. Opposition, in turn, diminishes the likelihood that the facility will be sited.
- A federal governmental versus a unitary structure lessens the probability of expressed opposition, and the direct effect of decentralized decision-making is to reduce the probability of siting.
- The analysis also suggests that the inclusion of mechanisms for public involvement in past siting programs has tended to occur in cases when there is expressed opposition, but such mechanisms have had no statistical effect on the outcome of past siting efforts.

As described in Section 3, the key findings from the study of the trend of aggregate public opinion over four decades, based on diverse questions regarding nuclear energy from multiple US nationwide surveys, include

- Widely known nuclear events, such as Three Mile Island (TMI), Chernobyl, and Fukushima, have substantial and sustained negative effects on the risk perceptions and acceptance of nuclear energy for residents of the US.
- These effects decay over time, but at different rates. Model estimates indicate that domestic nuclear crisis events like TMI have a dampening effect for approximately a decade. Events overseas, like the Chernobyl nuclear disaster in 1986, have a negative effect on nuclear attitudes lasting for roughly five years.
- Once the effects of specific nuclear events have been accounted for, our models indicate that there is an underlying decline in both perceived nuclear risks and the acceptability of nuclear energy. The rate of decline in perceived risks and nuclear acceptance has decreased over time, and may have reached a steady-state by 2011.

As described in Section 4, our time-series analysis of the content of social media analyzed (a) the content and volume of Twitter postings (tweets) and (b) Google searches that employed terms relevant to nuclear energy and nuclear waste management. These data allow analysis of shifts in public attention before, during and after major nuclear events like that in Fukushima, Japan following the March 2011 earthquake and tsunami. The key findings from this study include

- Public attention to both nuclear energy and nuclear waste management "spiked" immediately after the event.
- Attention declined approximately five weeks after the initial spike, but remained at significantly higher levels, roughly doubling the number of posts and information searches that had been made prior to the event.
- Both the Twitter and search data can be analyzed by location; the areas that experienced the largest increases in both kinds of indicators of attention were areas in which nuclear issues and facilities were present.
- The analysis of social media supports the analysis of the content of postings, such that issues of key importance to the public can be identified and addressed. This kind of information, evaluated over the course of a nuclear facility siting initiative, could provide important public input to programmatic and policy decisions.

Acronyms

AIC	Akaike Information Criterion (model aptness measure)
BIC	Bayesian Information Criterion (model aptness measure)
BRC	Blue Ribbon Commission on America's Nuclear Future
CRCM	Center for Risk and Crisis Management
DOE	Department of Energy
EPA	Environmental Protection Agency
GCC	Global Climate Change
IAEA	International Atomic Energy Agency
IP Address	Internet Protocol Address
LULU	Locally Unwanted Land Use
NAS	National Academy of Sciences
NEI	Nuclear Energy Institute
NGO	Nongovernmental Organization
NIMBY	Not in my Back Yard
NPP	Nuclear Power Plant
NRC	Nuclear Regulatory Commission
OLS	Ordinary Least Squares
PRIS	Power Reactor Information System
SD	Standard Deviation
SEM	Simultaneous Equation Model
SNL	Sandia National Laboratories
TMI	Three Mile Island
TVA	Tennessee Valley Authority
UNF	Used Nuclear Fuel
US	United States
WIPP	Waste Isolation Pilot Plant
YMP	Yucca Mountain Project
YV	Yankee Vermont

1. Introduction

The United States (US) program for siting interim storage and permanent disposal facilities for used nuclear fuel (UNF) is at a crossroads. The March 2010 request by the US Department of Energy (DOE) to the US Nuclear Regulatory Commission (NRC) for termination of the Yucca Mountain Project

(YMP) license application, followed one year later by the disastrous nuclear events in Fukushima, Japan, have resulted in a fundamental reconsideration of approaches for siting interim and permanent UNF management facilities in the US. The final report of the Blue Ribbon Commission on America's Nuclear Future (BRC) (BRC 2012) constituted a major milestone in that reconsideration. It called for abandoning the top-down, primarily technically driven facility siting approach outlined in the original Nuclear Waste Policy Act 1982 (and the subsequent Congressional selection of the resulting top-ranked Yucca Mountain site in the 1987 Amendments) in favor of a "new, consent-based siting approach to siting future nuclear waste management facilities" that is flexible and dependent on potential host communities, in collaboration with states and tribes, volunteering to be considered as candidates for choosing technically *and* socially acceptable sites.

In the DOE response to the BRC report, DOE endorsed the key principles of the BRC recommendations and proposed a strategy that "includes a phased, adaptive, and consent-based approach to siting and implementing a comprehensive management and disposal system" (DOE 2013: 1).[1] Hence, the BRC recommendation and DOE response constitutes a fundamental change in approach that may be considered by Congress in the future. This new process will be well served by a clear understanding of the trends, conditions, and program design elements that have shaped prior siting experience and that will influence public support for UNF facility siting in the future.

This report provides the results of a package of on-going social science studies undertaken by the Center for Risk and Crisis Management (CRCM) at the University of Oklahoma in collaboration with Sandia National Laboratories (SNL). These studies have been designed to test some of the widely held assumptions about the conditions under which siting does and does not work; to further understanding of when and how major nuclear events (like that in Fukushima, Japan) focus public attention and reshape public understanding and support for nuclear facilities; and to evaluate how the design features of siting programs can facilitate the legitimacy of and support for a siting program among the US public.

While the studies and results described here will be of broad interest to those involved in siting nuclear facilities, each of the studies focuses on different aspects of the problem and therefore may be of particular interest to individual readers. For those chiefly interested in public preferences for future siting efforts within the US, the survey results reported in Section 5 will be of chief interest. For readers interested in rigorous analysis of the global history of the outcomes of nuclear facility siting efforts, Section 2 will be of particular interest. Section 3 traces the trends in the US public's sense of the risk and acceptability of nuclear energy over the past four decades, with a central focus on the magnitude and duration of the effects on public opinion of the disastrous nuclear events at Three Mile Island (TMI), Chernobyl and Fukushima. Section 4 utilizes data from new sources – social media (Twitter) postings and Google search patterns – to trace the changes in public attention to nuclear issues before, during and after the Fukushima event.

In combination, the studies described in this report provide a broad, empirically grounded assessment of past nuclear facility siting efforts, the changing long-term patterns of public perceptions of the risks posed by nuclear energy in the US, and a detailed analysis of current American preferences for the design of fair and effective processes for siting UNF storage and disposal facilities. The research is informed by over two decades of experience in studies of the social and public policy aspects of nuclear programs by the research team, based at the University of Oklahoma[2].

The first study, described in Section 2, provides a first of its kind analysis of the historical pattern of success–and failure–in siting nuclear facilities (primarily nuclear reactors because of their prevalence). Using a global database of the siting initiatives for 269 nuclear facilities in 31 countries that have become either operational or were cancelled, the study permits quantitative modeling of some of the key factors that shape the probability that a facility will be become operational. The results of the model indicate that the most important factors conditioning siting success are structural – consisting of the openness and responsiveness of the political system to public (and opposition) input. Perhaps most sobering is the finding that, regardless of the nature of the institutional system (that is, the structure of the governing legal system and organizational allocation of authority) within which siting is taking place, there is a statistically significant long-term global trend in the direction of decreased probability of siting facilities. Additional findings, based on a subset of the siting data for which more extensive information was available, are that the addition of traditional mechanisms for public involvement (such as public hearings) have had little independent effect on probability of siting past facilities. It is important to note that these findings are based on historical data, and the trends and patterns that are described here led to the call for an overhaul of the UNF facility siting approach in the US by the BRC. A key contribution of this analysis is that, consistent with the BRC's recommendations, the conditions that influence siting outcomes are tested statistically using compiled data on efforts to site nuclear facilities, and the magnitude of the effects of key variables are estimated.

The second study, as described in Section 3, focuses more directly on the US experience, analyzing the long-term evolution of public preferences for nuclear energy and the response to major nuclear events such as the TMI, Chernobyl, and Fukushima accidents. The study employs a unique time-series dataset constructed from an array of indicators of public perceptions of the risks posed by nuclear energy, and support for nuclear energy, over the period from 1973-2011. These data are constructed using an innovative method for detecting larger underlying trends in public perceptions and preferences over past decades by combining an array of distinct but correlated indicators of "public mood" concerning nuclear issues. These data permit quantitative analysis of the history of how the US public has perceived the risks posed by nuclear energy, and their support for continued reliance on nuclear energy sources. The analysis of changes over time provides an empirical assessment of how historical events at nuclear installations have influenced public perceptions and preferences about nuclear energy, with direct implications for understanding the current post-Fukushima environment.

The third study, discussed in Section 4, examines how public attention shifts to (and from) nuclear issues, and how those changes can re-shape public concerns for the management of nuclear waste. For this study we employ two distinct kinds of real-time indicators of public attention: supply-based indicators, as measured by posted messages using social media, and demand-based indicators, as measured by the frequency of terms used in Internet web searches. The continuous feed and large volumes of these kinds of data streams permit analysis of changes in interest and attention on a moment-by-moment basis, such that both near-term and longer-term changes in attention are evident. It is also possible, with the social media data, to evaluate the content of the messages obtained in ways that indicate the directional change of public attitudes. We use these data to focus on the shifts in public attention that occurred with the onset of the crises at the Fukushima nuclear reactors, following the Tōhoku earthquake and subsequent tsunami that struck Japan on March 11, 2011. The analysis shows that both demand- and supply-based attention to issues associated with nuclear energy spiked shortly after the onset of the event, and that while attention declined following the event it has remained at a notably higher level than prior to the event. Attention to nuclear waste management also spiked; however, public attention settled to the same levels that had been evident prior to the event.

The final study reported in Section 5 is based on a nationwide Internet survey, conducted in June of 2012, of a sample of 2,017 adult respondents across the continental US. This survey is one in a series dating from 2006 that measures US public understanding and preferences on energy and environmental issues. The survey measured public beliefs and preferences, beginning with broad issues that will shape the context of energy policy debates and then focused on progressively more specific issues and choices concerning nuclear energy and nuclear waste policy. One of the innovations, made possible by using Internet survey methods, is that each respondent's location could be determined, which permitted experiments in which the respondent could be shown a map of their proximity to existing UNF storage facilities. When respondents are informed of their proximity – as would occur in the course of a local nuclear facility siting initiative – acceptance of a hypothetical interim storage or permanent disposal facility increases significantly. The survey then focused on public support for consent-based processes for the siting of UNF management facilities. Respondents provided insights into the characteristics of a consent-based siting process that would have the greatest support, including identification of which residents, officials and groups should have a say in siting, and at what stage within a consent-based process a community could no longer withdraw as a candidate site.

Although implementation of many aspects of the DOE proposed response to the BRC require Congressional action, DOE has undertaken activities within existing Congressional authorization, to be responsive to a near-term recommendation of the BRC that DOE learn as much as possible from prior experience related to facility siting. As stated by the BRC (BRC 2012: 118):

> To ensure that future siting efforts are informed by past experience, DOE should build a data base of the experience that has been gained and relevant documentation produced in efforts to site nuclear waste facilities, in the United States and abroad. This would include storage facility and repository siting efforts under the NWPA [Nuclear Waste Policy Act] by both DOE and the Nuclear Waste Negotiator.

The studies of this report add to the stock of knowledge that will facilitate "developing plans for consent- based siting processes" for UNF interim storage and permanent disposal facilities in the US (DOE 2013: 2).

2. Nuclear Facility Siting Cases: International Context

The history of efforts by industry and governments in many countries to site nuclear facilities provides an important opportunity to evaluate how the characteristics of nations and their governing institutions influence the implementation of nuclear facility siting programs over time and space. Beginning in the 1950s, policymakers around the world encouraged development of commercial reactors and utilities made hundreds of decisions about where nuclear power plants and/or storage facilities should be constructed. In many instances, utilities spent several years and millions (if not billions) of dollars on the development of these sites, only to see them cancelled before they became operational because of variety of factors. In other cases, the end result was an operable facility. In this study, we explore some of the factors (e.g., political, proximity in time to nuclear crises, and year of siting) that explain this variation. *Once a site has been identified for possible future use as a nuclear facility, what factors influence the probability that the proposed facility will become operable?*

Much of the current understanding of the variables that influence siting outcomes (i.e., whether a proposed facility is sited at a candidate location) is derived from case-based evaluations of nuclear facility siting efforts (e.g., the Waste Isolation Pilot Plant (WIPP) in New Mexico, or the Yucca Mountain repository site in Nevada). The case-based approach is of great importance, as recognized by the reliance on both the qualitative and quantitative analysis of individual cases in the BRC findings (BRC 2012), because case studies allow for the identification of the rich array of factors that may have influenced the course of the siting efforts in a specific context at a particular location. The challenge posed by the case approach, however, is that the number of variables available to explain the outcome of the siting exercise exceeds the number of cases, rendering hypothesis testing and the drawing of general conclusions impossible. Use of notable cases (or notable features of particular cases) as the basis for "lessons learned" poses the risk of overemphasizing particular variables that may have different effects in other cases. Some apparent relationships in a case will inevitably be stochastic (or ideocratic), and when they draw attention they may result in learning the wrong lessons. As is the case in science as well as public policy, "unlearning" wrong lessons can be costly and time consuming.

For these reasons it is important to supplement the use of case-specific studies with quantitative *comparative* analyses of larger sets of cases, permitting hypothesis testing and the accumulation of evidence about features of siting programs that systematically influence siting outcomes. It is important to remember that comparative studies also have their own limitations; the need for quantification across a large number of cases limits the subtlety and precision with which important variables can be operationalized, and some potentially important variables may be omitted altogether from the analysis because relevant documentation is unavailable or due to the absence of valid and reliable measures. In short, qualitative and quantitative analyses have different strengths and limitations, yet both are needed to provide the kind of cumulative knowledge base for effective facility siting that was called for in the BRC's final report (BRC 2102). In an extensive search for quantitative, comparative studies of this kind to date, however, we have found no such studies drawn from multiple cases of nuclear facility siting efforts.

In order to remedy this lack of comparative studies, this section describes an original comparative quantitative analysis of nuclear facility siting outcomes, using data derived from coding a large sample of cases (n= 269) to evaluate how the context of the siting program and elements of the program itself influence siting outcomes. We chose to focus on a few key factors: the level of democratic authority, the structure of the government system, the influence of well publicized nuclear disasters, the influence of opposition to the siting effort, and the role of public outreach programs.

One focus of this analysis is on the broad institutional configurations of the governmental systems of the country in which the siting takes place; does the overall "democraticness" of the country (at the time of the decision) change the probability that a proposed facility will be sited? Is a centralized and unitary system more likely to site a facility than one that decentralizes decision authority across layers in a federal-type system? Also of interest is the influence of key nuclear crisis events, such as the partial meltdown of the reactor at TMI in 1979, or the disastrous events at Chernobyl in 1986 and Fukushima in 2011. What is the magnitude of the effects of these kinds of crises on siting efforts? At the program level, what is the influence of opposition to the siting effort? Does the inclusion of public outreach programs improve prospects for siting a facility? In short, this analysis focuses on the way in which the broader institutional and political context, opposition and outreach have historically influenced the probability that a proposed nuclear facility was sited or abandoned.

2.1. Research Design, Cases, and Data

To develop the data for this analysis, we compiled a list of 269 known nuclear facilities, combining operational, decommissioned, and cancelled nuclear facility sites from the US and 31 other countries. (See Appendix A for a complete listing of cases.)[3] Note that due to information gaps, efforts to formulate a complete list of nuclear facilities proved to be quite challenging, especially for sites outside the US and in the earlier years of nuclear development. The difficulty in finding accurate data is particularly acute for cancelled, non-US sites (nuclear facilities that were firmly proposed by the governments but never became operational). Nevertheless, the list of cases compiled for this project covers a substantial number of the known operational, decommissioned, and cancelled nuclear fuel cycle facilities from the US and globally.[4]

Table 2-1. Percentage of Proposed Nuclear Facilities Completed by Country

Country	Total Cases	Percent Completed
United States	115	67.8
France	24	91.7
Germany	22	81.8
Japan	19	89.5
United Kingdom	16	87.5
Spain	10	70.0
India	7	85.7
Switzerland	6	66.7
Canada	5	100.0
Italy	5	80.0
Sweden	5	80.0
Ukraine	5	100.0
Other	29	90.7

The dataset includes a total of 115 US facility-siting efforts, and 154 international facility- siting efforts (excluding Russia, China, North Korea, and the former Soviet Union for reasons of the lack of access to relevant information). Of the 115 US observations, 78 are currently operational (or now decommissioned but once operational) facilities and 37 are siting efforts that were cancelled before they could become operational. Of the 154 international observations, 133 are currently operational or now decommissioned but once operational facilities, and 21 are siting efforts that were cancelled before becoming operational. A listing of the number of proposed facilities for which data were available, and the percentage of the identified facilities that were sited, is shown in Table 2-1.

Once the list of cases was identified, the data collection effort to characterize each case proceeded in a two-stage coding process. In the first step, all cases were evaluated and coded for a limited set of key country and context characteristics. In the second step, a randomly selected sub-set of cases were selected for more extensive analysis, in which information was coded concerning the social response to the siting effort and programmatic efforts for public outreach. Each of these steps, and the associated variables coded, are described in turn in the following sections.

2.1.1. Coding stage 1

In order to identify the primary country and context characteristics for the entire set of proposed nuclear facilities, we undertook a systematic search of documents and governmental websites, followed by an extensive Internet search for any information on operational, decommissioned, and cancelled cases. Usable documents were logged and archived. The information available for cancelled site cases proved to be much more difficult to obtain than was true for operational or decommissioned site cases. This may be because cancelled cases and siting efforts are less heavily reported, especially the ones that get withdrawn early in the siting process.[5] For all 269 cases, we coded the following variables for the year in which the siting decision was made to either open the facility or to abandon the siting effort: the host nation's polity score characterizing the degree of democratic openness to participation and dissent; the level of policy-making decentralization (or centralization) at the time of the decision; and the proximity in time to a major, globally reported, nuclear crisis. Each of these variables is described in turn.

a. **Polity Score (X_3):** This variable employs a widely respected measure of the characteristics of regime authority for a country on a scale measuring the relative democratic quality of authority from democratic to autocratic. The key elements of the polity scale are how executive leaders are recruited (e.g., elected by the public, appointed by a select group, or chosen on the basis of hereditary monarchies), the degree to which there are constraints on the exercise of executive authority, and the extent of open and legitimate competition for executive leadership.[6]

As shown in Table 2-2, some of the countries included in the siting database show substantial variation over time on the Polity scale, especially for countries such as Bulgaria, France, Germany, and South Korea. This is because the coding for the Polity variable is based on the final decision year regarding the facility—the year the site either became operational or was cancelled. For example, the polity scores for South Korea varied from -8 to +6 on the Polity scale due to the nature of the democratic quality at the time of the siting decision.[7]

Based on previous research on the nature of democratic processes, we expect that the outcomes of nuclear facility siting initiatives over time have been related to the characteristics of the host nation as captured by the Polity score. Given the long history of nuclear opposition (Weart 2012), we expect that opposition groups have been more effective in opposing siting in countries that are characterized by higher Polity scores, and less so in those on the lower (autocratic) end of the scale.

b. **Decentralization (X_2):** This variable measures the degree of political decentralization within a host siting country at the time of the decision to open or abandon a nuclear facility. The unitary form of government is one in which (a) the primary ruling power is held in the national or central government structures, (b) the government is highly centralized, and (c) states or local authorities do not hold autonomous political power. Federal forms of government, on the other hand, divide political power between the central government and the states, providing regional governments with some autonomy and influence in decision-making. Although the level of federalism can vary from country to country, this type of government provides an added layer of complexity in decision-making, greater access to the decision-making process by a range of interests (including those seeking to block the facility siting), and diffusion of the possible veto-points in the siting process. For the decentralization variable, federal (decentralized) systems were given an indicator variable[8] code of 1, and centralized systems coded as 0.

Table 2-2. Example Polity Scores by Country, Facility, and Year

Country	Facility	Decision Year	Polity Score
Bulgaria	Belene	2012	9
Bulgaria	Kozloduy	1974	-7
France	Le Carnet	1997	9
France	Plogoff	1981	8
France	Brennillis	1966	5
Germany (East)	Wurgassen	1971	-9
Germany (West)	Breisach	1973	10
South Korea	Kori	1977	-8
South Korea	Uljin	1988	6
South Korea	Wolseong	1982	-5

Based on previous research on the nature of political institutions and the influence of political centralization on policy outcomes, we expect that the siting of nuclear facilities will prove to have been more difficult in federal political systems as compared to unitary ones. Countries with a higher degree of decentralization allow for greater representation of interests through a diffused design of policy making and policy implementation. In contrast, unitary systems are characterized by concentration of power at the national level, leading to decreased opportunities for involvement at the state, provincial or local level.

c. **Proximity in time to major nuclear events (X_4):** This variable measures the proximity in time of the siting decision (to cancel or make operational) after a major nuclear accident such as TMI, Chernobyl, or Fukushima. Based on existing literature, we expect that nuclear facilities have been more difficult to site when preceded by a major nuclear crisis event. Furthermore, due to shifts in the public perceptions of nuclear risks in the aftermath of large-scale nuclear accidents, these kinds of nuclear events are likely to have increased the probability of opposition. For instance, in the case of the siting of SNR-300 Fast Breeder Reactor in Germany, sources indicate that public protests opposing the site "reached new heights" in the aftermath of the 1979 nuclear accident at TMI.[9] To code this proximity, each case was coded as having been proximate to a major nuclear event if the siting decision was made within 3 years following the TMI, Chernobyl or Fukushima events.[10]

d. **Year of Siting Decision (X_1).** It is plausible that the difficulty of siting facilities has changed over time, as suggested by historical accounts of nuclear facility siting cases (Weart 2012; Mahaffery 2009). For that reason, we utilize the date of the decision to operationalize (or abandon) the proposed nuclear facility to identify any underlying global trend in the pattern of siting outcomes. Given the historical accounts of growing difficulties in siting nuclear facilities, we expect to find a negative underlying trend in decisions to site nuclear facilities, once we have controlled for the effects of the periodic nuclear events.

In sum, the first stage of the data coding permit us to test hypotheses concerning the effect of major country characteristics – placement of the democratic/autocratic Polity score, and decentralization – along with measures of the effects of nuclear crisis events and underlying trends in time. As is evident in Table 2-1, the range of siting outcomes across the cases is sufficient (with siting percentages ranging from 67% to 100%) to provide a reliable empirical basis for testing these hypotheses concerning the primary determinants of nuclear facility siting outcomes.

2.1.2. Coding stage 2

While the Stage 1 coding enables clear hypothesis testing of important determinants of siting outcomes, it does not permit analysis of critical characteristics of the public response to the siting effort and programmatic efforts to address public concerns. Experience with the collection of case data demonstrated the challenges (in costs and time) associated with collecting detailed information on a large number of cases. At the same time, utilizing cases for analysis in which data collection was relatively "easy" posed the problem of serious selection bias for our modeling and hypothesis testing: if those cases for which data were readily available differed systematically from those for which it was difficult, the analysis would be biased and results unreliable. For that reason, we concentrated our resources on data collection for a more limited number of cases selected on a random basis form the larger data set.

From the list of 269 cases, we randomly generated a sub-sample of ~20% of the cases, providing us with a subset sample of 53 cases. (See Appendix A for a list of these cases.) Using this set of cases, we proceeded to code and collect data on two additional explanatory variables. This stage of the data collection process involved collecting additional data on the evidence of public opposition, and public outreach programs.

e. **Evidence of Opposition (X_6):** This variable captures the evidence of some form of opposition during the process of siting nuclear facilities. In other words, this variable includes information about whether or not the facility siting faced any opposition from nongovernmental organizations, the general public, and/or private companies. It is important to note that this variable does not capture the intensity or depth of opposition, but is simply an indicator variable of whether there was any documented evidence of social friction as a result of an attempted siting. Based on our research, 26 of the 53 cases in our Stage 2 sample had some accessible evidence of public opposition. We expect that nuclear facilities are more difficult to site when faced with opposition as compared to no opposition.
f. **Evidence of Public Outreach (X_5):** This indicator variable reflects evidence of formal mechanisms for public involvement employed in the siting program during the process of siting nuclear facilities. In other words, this variable codes whether or not the operating entity or the government policy design incorporated means of engaging the public or opposing groups in any way. The overwhelming majority of cases for which such mechanisms were evident included formal hearings or other means of informing the public, followed by mechanisms for public comment and expression of opinion. In most instances, descriptions of any public outreach effort were minimal, such that we were unable to code for variation in these efforts. Based on our data search, 23 of the 53 cases in our sample had some evidence of public engagement. Our expectation was that nuclear facilities would have been more difficult to site without measures for public engagement.

The coding scales and distributions for the dependent and explanatory variables are presented in Table 2- 3. We applied logistic regression to model the relationship between siting outcomes and the primary institutional variables from the Stage 1 data (Table 2-4).[11] More specifically, we modeled the probability of nuclear facility siting in stages, first testing for a long term-trend in siting outcomes (Model 1), then adding the national-level decision-making characteristics (Model 2), and finally the effects of proximity in time to a major nuclear crisis event (Model 3).

Table 2-3. Scales for Stage 1 and Stage 2 Coding and Distributions for the Dependent Siting Outcome Variable and the Explanatory Variables

Variables	Minimum	Maximum	Mean	Number of Cases (n)
Siting Outcome (binary) (Y_1)	0	1	0.784	269
Decentralization (binary) (X_2)	0	1	0.606	269
Democracy (Polity Score) (X_3)	-9	10	8.164	269
Nuclear Crisis Event (binary) (X_4)	0	1	0.242	269
Year of Siting Decision (X_1)	1956=0	2012=56	23.152	269
Public Engagement (binary) (X_5)	0	1	0.434	53
Public Opposition (binary) (X_6)	0	1	0.491	53

2.2. Analysis and Findings

Table 2-4. Coefficients for Explanatory Variables of Three Logistic Models of Siting Outcome from Stage 1 Coding

	Coefficients for Siting Outcome Model ($E\{Y\}$)		
Variable Coefficient	**Model 1**	**Model 2**	**Model 3**
β_0 (Intercept)	3.259*(0.452)[a]	11.167*(3.637)	11.826*(3.666)
Year of Siting Decision (1956=Year 1) β_1	-0.078*(0.016)	-0.099*(0.019)	-0.089*(0.019)
Decentralization β_2		-0.994*(0.429)	-0.842^{+}(0.431)
Democracy (Polity) β_3		-0.693*(0.359)	-0.768*(0.364)
Nuclear Crisis Event β_4			-1.013*(0.366)
Number of Cases n	269	269	269
AIC[b]	256.101	231.195	225.561
BIC[c]	263.290	245.573	243.535
Likelihood-ratio[d]	28.358	57.265	64.898

* Coefficient significant at $p < 0.05$ in a one-tailed Wald test.

+ $p < 0.1$ in a one-tailed Wald test.

[a] Standard error.

[b] Akaike Information Criterion (AIC), relative measure of model aptness, value decreases as model aptness for forecasting improves.

[c] Bayesian Information Criterion (BIC), relative measure of model aptness, value decreases as model aptness for forecasting improves.

[d] Ratio for testing whether variables may be dropped in Likelihood Ratio Test.

Table 2-4 lists the coefficients and standard errors that we derived for modeling the probability of nuclear facility siting using the Stage 1 data. Note that the effects of the independent variables are statistically significant at the 0.05-level, with the exception of decentralization (which was significant at the 0.1- level). The estimated effects of the independent variables were all in the hypothesized direction. Also note that the addition of variables to model improves model fit (as shown by the declining BIC and AIC model aptness values).

Focusing on Model 1, the probability of completed nuclear siting declines substantially over time. Transformation of the log-odds to probabilities shows the probability of siting to decline over the 56-year time-span of our series by 0.52. Second, as shown in Model 2, more democratic institutions (as indicated by their Polity scores) are associated with more difficulty in nuclear facility siting; decentralization through federal systems has modestly reduced the probability that a facility will be sited.

Table 2-5. Coefficients for Explanatory Variables of Logistic Models of Siting Outcome and Public Opposition from Stage 1 and Stage 2 Coding

Variable Coefficients	Direct Effects Siting Outcome (Y_1) Model 4	Indirect Effects Public Opposition (X_6)
β_0 (Intercept)	3.556*(1.026)	-13.893 (9.171)
Decentralization β_0	-1.772* (0.924)	-2.704*(1.743)
Public Engagement β_5	0.548 (1.158)	5.549*(1.734)
Nuclear Crisis Event β_5	-0.974*(0.787)	3.334*(1.730)
Public Opposition β_6	-1.927*(1.193)	
Democracy (Polity Score) β_5		1.288^{+}(0.940)
Number of Cases n	53	53
AIC[b]	57.086	39.884
BIC[c]	66.938	49.735
Likelihood-ratio[d]	11.966	43.571

[*] Coefficient significant at $p < 0.05$ in a one-tailed Wald test.

+ $p < 0.1$ in a one-tailed Wald test.

[a] Standard error in parentheses.

[b] Akaike Information Criterion (AIC), relative measure of model aptness, value decreases as model aptness for forecasting improves.

[c] Bayesian Information Criterion (BIC), relative measure of model aptness, value decreases as model aptness for forecasting improves.

[d] Ratio for testing whether variables may be dropped in Likelihood Ratio Test.

All else equal, the predicted difference in the probability that a facility will become operational in a country that is high on the Polity scale (10) versus one that is low on the Polity scale (-9) is -0.21. Similarly, the probability that a site will become operational in a decentralized country is 0.04 lower than in a centralized country. Finally, as shown by Model 3 of Table 2-4, the effect of proximity in time to a nuclear crisis event was to significantly reduce the probability of siting the facility, with a change in probability of 0.07.

Table 2-5 shows the model results when predicting siting outcomes, adding the measures indicating public opposition and public outreach.[12] As shown for Model 4 of Table 2-5, we used logistic regression to ascertain the direct effects of crisis events, decentralization, opposition, and engagement on the probability of siting nuclear facilities. In the third column of Table 2-5 we show the results of a another logistic regression model to show the effects of crisis, Polity score, decentralization, and public engagement on public opposition, which in turn, indirectly influences siting outcomes as shown in Model 4. Not surprisingly, the evidence of public opposition was fully explained by variation in the Polity score; apparently autocracies have little tolerance for opposition to nuclear facility siting.

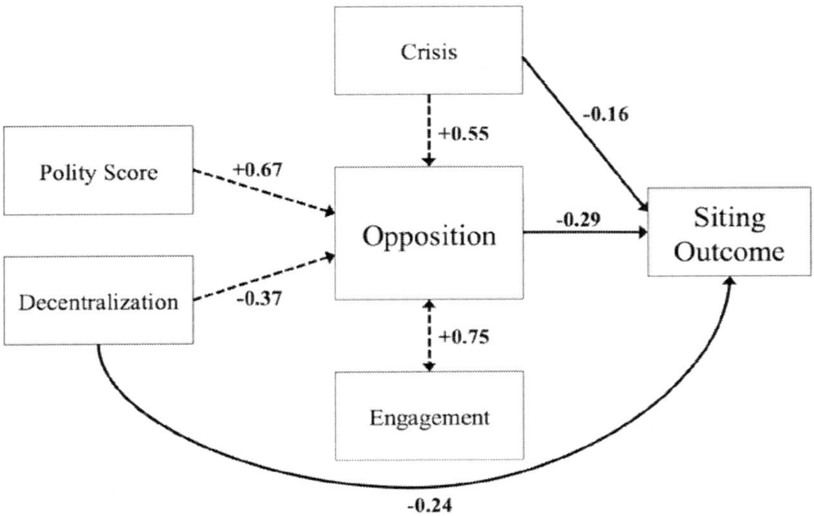

Figure 2-1. SEM Model Results for Nuclear Facility Opposition and Siting.

To facilitate interpretation of the model results, Figure 2-1 (below) presents these estimates of a simultaneous equation model (SEM), expressed in terms of the predicted probabilities associated with shifts (from low to high) in the explanatory variables. The solid lines in the figure indicate the statistically significant direct effects of decentralization, crisis, and opposition on the siting outcome in each case. The dashed lines in the figure indicate the statistically significant indirect effects of Polity score, decentralization, crisis, and engagement on the siting outcome in each case (as mediated by opposition).

First we focus on the model prediction of the probability of opposition to facility siting. The most potent predictor of opposition is the democratic character of the siting nation: transition from lower to higher Polity scores increases the probability of opposition. All else equal, the probability of opposition in a country at the low end of the Polity scale 0.67 less than is the probability of opposition at the high end of the scale. Decentralization appears to have the opposite effect—the probability of opposition in decentralized (federal) countries is 0.37 less than is opposition in unitary systems of government. We also find that a crisis increases the probability of opposition. Holding the other variables at their mean, the predicted probability of opposition is 0.55 more likely when the siting decision was preceded (by 3 years or less) by the TMI, Chernobyl, or Fukushima events. Finally, after controlling for decentralization, crisis, and Polity scores, there is a somewhat surprising *positive* association between public outreach efforts and opposition. When public outreach is evident, the probability of opposition is 0.75 greater (and vice versa). This suggests a more complex relationship between opposition and outreach. Because of this complexity, the arrow between opposition and outreach in Figure 2-1 is shown double-headed, rather than uni-directional.

Turning to the direct effects on siting outcomes shown in Figure 2-1, facility siting is estimated to be more difficult (0.24 less probable) in decentralized/federal as opposed to centralized/unitary systems. Crises also make siting even more difficult. If a nuclear crisis event preceded the siting effort by 3 years or less, the probability that a site became operable is 0.16 lower than otherwise. Opposition is also a potent predictor of facility siting outcomes; the predicted probability that a site will become operable absent opposition is 0.29 lower when there is evidence of opposition. Lastly, when controlling for federalism, crisis, and opposition, there is no discernible relationship between programmatic public outreach and siting outcomes.

2.3. Summary and Implications

The comparative analysis of nuclear facility siting cases, coded for key characteristics of the siting country, the siting context, and the program gives us the opportunity to empirically test conjectures about what increases or decreases the probability that a proposed facility will, in the end, be completed. This analysis, in compiling and coding 269 siting cases from plans to site nuclear facilities in 31 different countries over six decades, permits tests of a number of these hypotheses, as shown above.

One of our key findings is that, all else being equal, the probability that a proposed nuclear facility will become operational has decreased substantially over time, from near certainty in the mid-1950s to a fifty- fifty proposition at the present time. Our analysis also indicates that variation in the institutional frameworks for decision-making within countries explains a substantial fraction of the differences in siting outcomes. More democratic countries, and those with federal (decentralized) decision-making structures, face larger hurdles and lower likelihoods for nuclear facility siting than countries that are less democratic and more centralized. These are features of the siting context that are largely fixed within a country, but they frame the conditions within which proposed nuclear facilities and siting programs must be designed and implemented. While these conditions have long been the subject of conjecture, this analysis provides solid empirical evidence supporting those conjectures.

Why do countries with greater democratic openness, as defined by the widely used Polity scores, have a more difficult time siting nuclear facilities? A more detailed analysis of a randomly selected sub-set of siting cases provides some answers. Historically, greater democratic openness is associated with an increased probability of expressed opposition to the facility. Opposition, in turn, diminishes the likelihood that the facility will be sited. Federalism, on the other hand, appears to lessen the probability of expressed opposition, but the direct effect of decentralized decision-making is to reduce the probability of siting. Finally, in the record of cases used for this analysis, the inclusion of mechanisms for public involvement in siting programs is highly associated with expressed opposition, but it has no bearing of the likely outcome of the siting effort. Our findings also clearly emphasize the sobering finding that the probability of "success" for any particular proposed site had declined to only a 50:50 proposition over time, before changes were instituted in the international community.

These findings indicate that, as historically practiced, nuclear facility siting initiatives have been most effective when the public has had *less* opportunity to participate. This kind of context is most consistent with the top-down approach to siting described (and rejected) by the BRC. That approach, while successful in less open and democratic systems, has not fared well over time, as governmental systems have on average become more open and responsive to public concerns. When the context has afforded greater public voice, public opposition has been more prominent. Given the persistent evidence of fears of nuclear technologies (Weart 2012; Slovic et al. 1990), it is not surprising that opposition to nuclear facilities is more likely when public expression and participation is the norm. Because siting agencies continued to employ the kind of top-down facility-siting strategies for a number of years (a process that made sense in the context of less open and participatory systems), siting failures were increasingly common. In sum, new approaches and strategies were needed in the context of increasing participatory opportunities.

Why had the inclusion of public outreach programs with the old siting methods not increased the probability that a proposed facility would become operational? While the model results provide associations, rather than clear causal explanations, it appears to us that public outreach programs of the kind traditionally employed in nuclear facility siting efforts had been largely responsive to the expression of opposition once a site has been identified for a proposed facility.

In the past, when public outreach was implemented to address already-expressed opposition, it had no ability to increase the probability that a facility would become operational. This may be because the siting agencies continued to use siting strategies learned under a different context, and those strategies simply did not fare well in the world beyond the 1970s.

All of these findings are consistent with the conclusions of the BRC that in the US, in order to site nuclear facilities, it will be necessary to alter the traditional mode of the facility siting process from top-down designs (which in many cases was characterized as "decide, announce and defend") to ones that rely on consent-based programs that give host communities and states a prominent role.

Recently, broad engagement programs implemented early in the siting process been a part of the facility siting effort, but no cases have yet resulted in operational facilities that could be included in our analysis.

3. LONG-TERM CHANGES IN PUBLIC SUPPORT FOR NUCLEAR ENERGY IN THE US: 1973-2011

Section 2 of this report provided a quantitative analysis of the outcomes of global nuclear facility siting efforts, and found that (1) the major nuclear crisis events at TMI, Chernobyl and Fukushima increased levels of expressed opposition to proposed nuclear facilities, and (2) proposed facilities are statistically significantly less likely to become operational in the period following such events. The implication is that nuclear crisis events have potent, lasting negative effects on public support for nuclear facilities, and therefore on the prospects for siting such facilities.

This section narrows the focus to public attitudes about nuclear issues in the US, developing a new set of measures of public opinion over time that show the trends and variations over nearly four decades. These data permit statistical analysis of the responses of the US adult population following nuclear crisis events over ~40 years, allowing us to analyze how consistent and durable the effects of these crises have been. It also permits an assessment of the underlying trend in public attitudes toward nuclear energy in the US over the last four decades, separating these trends from the effects of the nuclear events. The intent is to enhance our understanding of public responses to nuclear events, and to distinguish those responses from the more general evolution of attitudes toward nuclear issues over time.

3.1. Introduction

The planning, construction, and operation of a nuclear facility is a long-term process that unfolds over many years and involves technology that is widely perceived to pose significant risks. This long time scale combined with the dynamic nature of public opinion and the perception of risk associated with nuclear technology means that nuclear siting necessarily takes place in a context in which substantial fluctuations in public support for nuclear technologies can occur (Jenkins-Smith et al. 2011). These fluctuations can result in periods in which public support for nuclear energy options is relatively positive – as will be shown to be the case in the recent decade-long period of nuclear optimism dubbed the "nuclear renaissance" – as well as periods of growing reluctance and opposition. Understanding those fluctuations, and how they respond to nuclear crises such as the recent events in Fukushima, can facilitate understanding of prospects for nuclear policy options and assist in the design of public engagement programs.

In this section we continue to re-examine the quantitative record of the evolution of public opinion about nuclear energy over the last four decades, which was started last year. While numerous surveys have included measures of nuclear attitudes, the consistency in question wording, sample size and periodicity in data collection has made long-term analysis of these data difficult. In this study we address this difficulty by combining data from multiple public opinion surveys over multiple years into consistent aggregated measures, enabling systematic empirical analysis of the development of public opinion. Using this technique we developed two over-time measures of public views of nuclear issues: (1) nuclear risk, which captures expression of perceptions of risk regarding nuclear energy, and (2) nuclear acceptability, which aggregates expressed preferences about acceptance of nuclear energy as an energy source. We then analyze the changes in nuclear risk and acceptability over time, and then following three of the most prominent nuclear accidents: TMI, Chernobyl, and Fukushima.

3.2. Study Design, Data, and Implementation

3.2.1. Approach for Measuring Public Preference Over Multiple Surveys

Given the importance of questions about the linkage between public policy and public opinion, scholars have developed sophisticated ways to measure aggregate opinion over time. One of the more prominent approaches concerns the study of "policy mood" as developed by political scientist James Stimson (Stimson 1999). Stimson notes that "mood" represents "general dispositions" toward domestic public policies that can be understood to underlie such concepts as "liberal" or "conservative" dispositions (Stimson 1999: 20). Stimson utilized answers to multiple survey questions over multiple years and across multiple policy issues to calculate an aggregate measure for this kind of "mood." His approach is to code survey responses as indicative of an underlying liberal to conservative attitudinal scale, including such disparate questions as perceptions of the performance of political institutions (Congress, the President, and the Courts), policy preferences, and more direct measures of ideological leanings (Erikson et al. 2002; Stimson et al. 1995). More recent work has extended this approach to the examination of issue-specific "moods," using Stimson's algorithm to estimate underlying dispositions with regard to a single policy issue (Atkinson et al. 2011a, 2011b). For example, a recent study used Stimson's algorithm to test hypotheses about drivers of climate change risk perceptions using multiple survey questions from 2002 to 2010 (Brulle et al. 2012).

We applied this approach to examine opinions about nuclear energy from 1973 to 2011. In order to gather a significant number of survey questions and responses, we used the iPoll database managed by the Roper Center, housed at the University of Connecticut.[13] To identify questions related to opinions about nuclear energy we used, "atomic energy," "atomic power," "nuclear energy," and "nuclear power" as search terms, yielding about 2,000 questions from 1945 to 2012. Following Stimson's procedure, we downloaded questions and "top-line" survey information including responses (which are coded as percentages, e.g., 51% support more plants, 49% oppose), sample size (number of respondents to the survey), and the dates in which the survey was administrated. We then identified those survey questions that were asked in more than one year, and the span of years for which adequate combinations of questions were available. Such a span was available for the years to 1973 to 2011. We then sorted the survey question into categories, such as those concerning the addition of more nuclear plants, which included very similar question content and wording. Once these categories were identified, the associated survey information for each question was entered into an analysis database. Once the survey information was collected, we used the WCALC computer program to calculate nuclear opinion indices using Stimson's algorithm.[14] In the next section, we briefly describe how the algorithm works.

3.2.2. Estimating Public Preference About Nuclear Energy

As noted, multiple survey questions taken over time and by multiple polling firms can be combined to estimate the underlying latent trend in public opinion. Using the process described above, we collected questions about nuclear energy and placed those into several question categories. However, the inconsistent way in which various polling firms administer surveys (e.g., different years, sample sizes) makes the straightforward aggregation of data, such as average per year, impossible. The Stimson algorithm accounts for such inconsistencies by standardizing responses using a common metric and by using a recursive process to estimate missing values. In short, the algorithm estimates opinion as the average of the available questions in time period t, weighed by the degree of shared variance of each question with the latent dimension. The algorithm is expressed in the form:

where

I = 1, *n* is all available question categories for period *t*

J = 1, *t* is all available comparisons for question categories *i*

b is the base period for the recursive metric

is the value of the metric for period *b*

is an estimated weight for the common variance of question responses, and opinion

This calculates aggregate opinion as a weighted moving average of past ( ) and future values ( ) of the survey questions. Where, the final period of the opinion index is set at an arbitrary value of 100 and then used to calculate missing values. Therefore, question category$_i$ at t-k is equal to: question category$_{i,t-k}$ = 100 * (question category$_{i,t}$ / question category$_{i,t-k}$).

The WCALC program implements the Stimson mood algorithm. Four pieces of information from each survey question are used by the WCALC program: a variable name (a truncated descriptor based on the question category, such as more plants, which represents all questions about the construction of more nuclear plants), the month/date/year of each survey question, a value representing the ratio of survey responses for the category, e.g., ratio of support or agreement to opposition or disagreement), and the sample size of the survey in which the question was asked. The WCALC algorithm "utilizes the relationships between survey [question details] to develop a measure of the central tendency across a number of different surveys over time" (Brulle et al. 2012: 4). Again, the Stimson algorithm assumes that all related questions are indicators of a more general attitude, and uses the data from the surveys to estimate that more general attitude over time. In addition, it imputes missing data based on past and future responses and weights responses by their relative sample sizes. The resulting algorithm scales were coded such that values for each year ranged from -50 to 50, where a value 0 reflects equal support and opposition.

3.3. Trend in Perceived Nuclear Risk

Using the survey data and the WCALC program, we constructed two nuclear opinion scales. The first scale concerns risk perceptions related to nuclear energy. Based on the iPoll search, the following categories of survey questions were created: those involving nuclear accidents, nuclear safety, nuclear risks, "not in my back yard" (NIMBY) safety, and NIMBY danger. These categories are based on similar question wording. For example, the nuclear accident category consists of questions related to the danger of nuclear accidents and concern about nuclear accidents. A total of 54 questions were sorted across these categories. The exact question wording, and the listing of survey sources and years in which the surveys were fielded, are organized by question category and included in Appendix B. Questions were coded so that a higher value indicated greater perceived risk. Table 3-1 shows the name of each of the categories of questions, the number of questions included, the number of years for which questions were available, and the correlation with the underlying risk dimension as calculated by the WCALC algorithm.

The perceived risk scale is coded from -50 (lowest level of perceived risk) to +50 (highest level of perceived risk). Figure 3-1 plots the risk scale over the 1973 to 2011 period.

Table 3-1. Components of the Nuclear Risk Scale

Category	Number of Questions	Number of Years	Correlation
Nuclear Accident	5	5	0.967
Nuclear Safety	26	14	0.767
Nuclear Risk	5	4	0.983
NIMBY Safety	10	10	0.974
NIMBY Danger	8	6	0.908

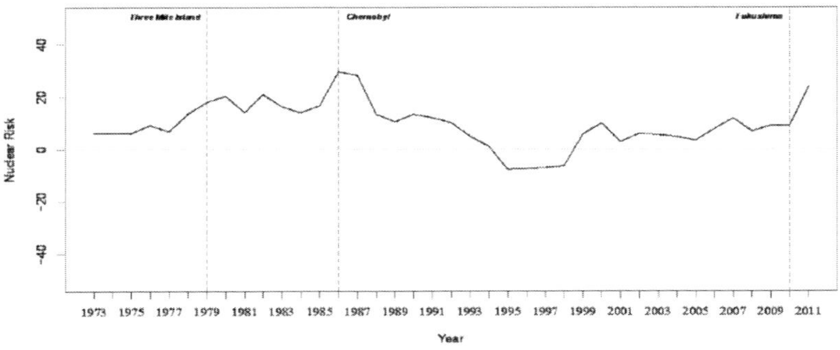

Figure 3-1. Perceived Nuclear Risk: 1973–2011.

As is evident in Figure 3-1, perceived risk grew from 1973 through 1980, and leveled off until a substantial jump was evident in 1986. Subsequently, perceived nuclear risk declined through the 1990s, reaching a nadir in 1995-1998. After a rise in 1999-2000, perceived risk remained stable through 2010.

In 2011, at the time of the Fukushima nuclear event, perceived risk rose steeply, as compared to the 1988 to 2010 period.

3.4. Trend in Nuclear Acceptability

The nuclear acceptability scale consisted of questions that were sorted into four broad categories including constructing more nuclear plants (more plants), the use of nuclear energy to produce electricity (nuclear electricity), increasing the use of nuclear energy (nuclear increase), and government promotion of nuclear energy (government promote). Overall, 233 questions were utilized for the coding, spanning 1973-2011. The exact question wording, and the listing of survey sources and years in which the surveys were fielded, are organized by question category and included in Appendix B. For each question a ratio was calculated based on the distribution of responses, and coded such that a higher value would indicate more support for nuclear energy.[15] The WCALC program was again used to generate the annual values for the series, based on the inter-item correlations over time. Table 3-2 shows the names of the question categories used to produce the series, the number of questions in each category, the number of distinct years in which questions representing that category were asked, and the correlation of the variable generated by that category of questions with the underlying latent dimension. See Appendix B for more details about the survey questions used.

Table 3-2. Components of the Nuclear Acceptability Scale

Category	Number of Questions	Number of Years	Correlation
More Plants	166	29	0.962
Nuclear Electricity	45	22	0.884
Nuclear Increase	11	8	0.904
Government Promote	11	6	0.936

Figure 3-2 plots the nuclear acceptability scale, and includes demarcations for each of the major nuclear incidents. Note that nuclear acceptability peaked in 1977, and then declined until 1990. Acceptability then gradually rose again, with some fluctuations until 2010. In 2011, the year of the Fukushima nuclear event, nuclear acceptance dropped. Although developed in a more sophisticated manner, the trend in this figure is in agreement with the analysis conducted last year (Jenkins-Smith et al. 2011, Figure 3-1). However, here the return to general acceptability occurs between 1992 and 1995, while the analysis last year did not show a return to general acceptability until after 2000.

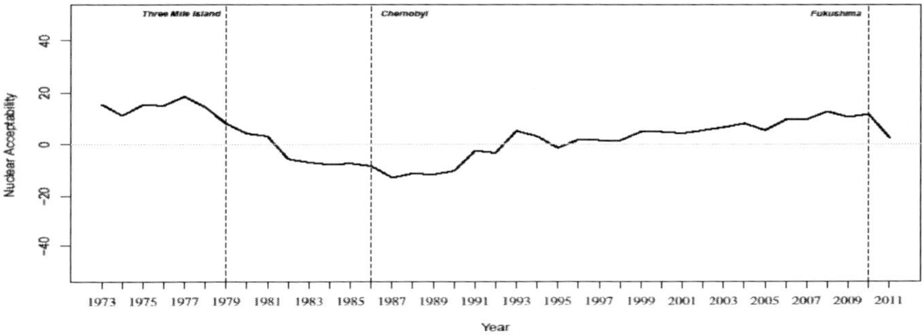

Figure 3-2. Nuclear Acceptability: 1973–2011.

3.5. Analysis of Nuclear Attitudes

While both of the nuclear attitude trends appear to be related to nuclear events, visual inspection does not distinguish between overall variation in the nuclear attitude trends and significant departures from the trend associated with events at particular points in time. To test for the effect of events, and to distinguish those effects from underlying trends over time, we used ordinary least squares (OLS) analysis with a measure intended to capture change over time.

Our first OLS model predicts changes in the risk scale. The OLS model is expressed as:

Y is nuclear risk and we used indicator variables for the periods of the TMI, Chernobyl, and Fukushima nuclear events. For the year variable, we included both in linear and quadratic form to capture changes in the underlying trend in nuclear risk perception. We were also interested in how long the impacts of the nuclear events might last, therefore we experimented with different decay periods for the effects of nuclear events on public attitudes, with time spans ranging from 3 years to indefinite (through the end of the series). We then assessed model fit for each model using the various time spans. The model that produced the best fit, determined by comparing the *F* statistic, adjusted R^2, and the residual standard deviations for each model, models a 10 year impact for TMI and a 5 year impact for Chernobyl.[16] Therefore we modeled the effect of TMI as lasting 10 years and the effect of Chernobyl as lasting 5 years from the onset of the crisis. The model results for the nuclear risk series is shown in Table 3-3.

As is evident in Table 3-3, each of the three nuclear crisis events is estimated to have had the effect of *increasing* perceived nuclear risk (sometimes described as risk ratcheting). The TMI event is estimated to have increased perceived risks by ~13% ($p<0.001$), while the Chernobyl event increased perceived risk by 9% ($p<0.01$). Finally, Fukushima is also associated with an increase in risk of nearly 14% ($p<0.05$).

Although the previous paragraph might imply that risk perception only increases, this is clearly not the case as shown in Figure 3-1. Hence, also of interest is the estimated trend in perceived risks over time, including the effects of the nuclear events. The first order effect is negative (-0.88 units per year, $p<0.05$), though the rate of decline in perceived risks decreases substantially over the series (as shown by the small but positive and significant effect of the square of the year variable). In the first year, the decline in the underlying trend in perceived risks is estimated to be -0.86%; by the mid-scale period (17th year) the decline in perceived risks was estimated to be only 0.54; and by the end of the series (in year 38) change in perceived risk was estimated to be 0.12 (a 0.14% *increase*).

In sum, perceived risks did eventually decline, but at a slowing rate over the span of the data. Within that period, the nuclear crisis events associated with TMI, Chernobyl and (apparently) Fukushima punctuated the longer-term perceived risk trend with upward spikes, the effects of which lasted on the order of a decade for TMI and five years for Chernoybl.

We conducted a similar statistical analysis on the trend over time in public views on the acceptability of nuclear power, though in this instance we included the annual perceived nuclear risk measure as one of the predictors of the measure of nuclear acceptance. The second OLS model is expressed as:

In this model, Y is nuclear acceptability and as with the perceived nuclear risk models, alternative model formulations showed that a ten-year lag effect for TMI and a five-year lag for Chernobyl best fit the available data. The underlying trend in nuclear acceptance was again modeled as a non-linear (polynomial) trend. The resulting model estimates are shown in Table 3-4.

Table 3-3. Estimated Coefficient of the Determinants of Perceived Nuclear Risk from Ordinary Least Square Regression

	Estimated Coefficients
(Intercept)	10.23***(2.58)
Three Mile Island	12.74***(2.30)
Chernobyl	9.19**(2.82)
Fukushima	13.70*(6.02)
Year	-0.88*(0.33)
$Year^2$	0.02*(0.01)
n	39
adj. R^2	0.63
Resid. sd	5.28

* Significant at $p<0.05$; ** $p<0.01$; *** $p<0.001$.
Standard errors in parentheses.

Table 3-4. Ordinary Least Squares Estimates of the Determinants of Nuclear Acceptance

	Estimated Coefficients
(Intercept)	18.38***(2.04)
Three Mile Island	-7.13**(2.09)
Chernobyl	-8.68***(2.11)
Fukushima	-13.46**(4.23)
Nuclear Risk	0.02 (0.11)
Year	-1.80***(0.24)
Year2	0.05***(0.01)
n	39
adj. R2	0.84
Resid. sd	3.45

* Significant at $p<0.05$; ** $p<0.01$; *** $p<0.001$.
Standard errors in parentheses.

As was the case with the trend in perceived nuclear risk, each of the nuclear crisis events was estimated to have a statistically significant effect on public acceptance of nuclear energy. The effects of the TMI event is estimated to have reduced nuclear acceptance by about 7% ($p<0.01$); the Chernobyl event by nearly 9% ($p<0.001$); and the Fukushima event by over 13% ($p<0.01$). The underlying trend in acceptability is downward, but at a decreasing rate over the time span of this analysis. In the initial year of the series, public acceptance declined 1.76 points (on the 100-point scale); by the series mid-point the decline was reduced to 1.12 points per year; and by the end of the series the estimated annual change was a positive 0.1 points per year (or near zero). Note that, having already accounted for the changes due to the nuclear events, perceived nuclear risk had no statistically discernible effect on nuclear acceptance.

Nuclear acceptance, like perceived nuclear risk, is responsive to nuclear crisis events. Each crisis diminished nuclear acceptance, and did so for sustained, though varying, periods of time. Note that, by our estimates, the dampening effect of crises on public acceptance is cumulative; the effect of one crisis effectively stacks on top of that of another. But having accounted for the effect of the nuclear events, there is still an evident underlying trend toward diminishing nuclear acceptance. By this account, the apparent rise in nuclear acceptance, beginning in 1990 but accelerating over the period from 1995 to 2010 (as shown in Figure 3-2), represents a reduction in the lagged effect of prior nuclear crises, rather than an underlying trend toward greater nuclear acceptance *per se*. Indeed, the rise in acceptance evident in 1990 occurs as the effect of the Chernobyl event (as estimated in our models) was wearing away.

What are the possible effects of the Fukushima event on future nuclear acceptance? To put the question differently, we can ask: How would nuclear acceptability have evolved in the absence of the Fukushima event? To make such an estimate, we ran a forecast model predicting what nuclear acceptability might have been without the effect of the Fukushima event. The forecast is based on a simple autoregressive model using only the last period (i.e., AR(1) model).

The 2010 and 2011 periods are omitted in the model in order to evaluate the estimated difference expected in 2010 (prior to the Fukushima event) and 2011 (the year of the event). This model uses the acceptability in the previous period (A(t-1)) to make a prediction about the acceptability in the next period (A(t)). Using the scale from 1973 to 2009 (two years prior to Fukushima), we predicted possible scores from 2010 to 2014.

This simple forecast model predicts a gradual decline in public acceptability, consistent with the estimated underlying nuclear trend noted above. The forecast model predicted nuclear acceptability to be +11.00 in 2010, while the observed score was +11.62. By 2011, however, the predicted score was +10.42, nearly 8 points above the observed score of +2.56. This illustrates the sizable negative shift in nuclear opinion induced by the Fukushima event, even when an underlying mild declining trend is included. We caution that a more refined assessment of the implications of the

Fukushima event will require more years of data and more extensive modeling.

3.6. Summary

This analysis has employed evidence from a very large set of surveys taken over time, in combination with an innovative data aggregation technique, to statistically analyze fluctuations and trends in public opinion over time regarding nuclear issues. Two data series – representing perceived nuclear risks and the acceptability of nuclear power – were constructed and analyzed. The observable patterns show substantial variation over time, and the series are consistent with recent experience. Statistical analysis demonstrates the potent effect of nuclear crisis events on public opinion. The occurrence of the TMI, Chernobyl and Fukushima events are all associated with increasing perceived nuclear risk and decreasing public acceptability. The data also suggest that the effects of these events are durable, lasting about a decade for TMI, and about half that time for Chernobyl before the effect decays. This discrepancy in duration may be a result of the fact that TMI occurred within the US, whereas Chernobyl occurred outside the US and within the former Soviet Union. The public may be reacting more strongly to events that are "closer to home." This finding could have important implications for the lasting impact of the Fukushima event on the American public's support for nuclear energy, but more time and more data are needed. However, we can conclude that these kinds of events clearly impose sizable and substantive declines in support for nuclear energy.

The analysis also shows that, apart from the event-induced fluctuations, there are significant and interesting trends in public opinion concerning nuclear issues. According to the models, perceived nuclear risks have trended downward since the mid-1970s, but at a declining rate. Over the same period, nuclear acceptability has also declined. It appears that perceived risk, though a major ingredient shaping the level of acceptance of nuclear technologies is not the only one.[17] This result is also consistent with our finding in Section 2 that the ration of outcomes of nuclear siting initiatives has shifted over time, with a growing percentage of proposed facilities being cancelled before the facility becomes operational. In combination these results reinforce the conclusion reached by the BRC that the impediments garnering public support for nuclear energy – and for the siting of the necessary facilities – are significant and new approaches are necessary.

4. Real-Time Indicators of Public Attention to Nuclear Issues

4.1. Introduction

Section 3 of this report focused on the evolution of US public opinion in the aftermath of nuclear crisis events. The analysis in this section turns to the related but distinct concept of shifts in *public attention;* by this we mean the relative attentiveness of the public to particular issues like nuclear energy and nuclear waste management. The analysis of public attention can provide policymakers and decision makers with feedback and/or signal that the public (or specific segments of the public) is concerned or worried about a particular issue and therefore more likely to be interested and attuned to the debate and decisions that are being made with respect to that issue.

Research in political science and public policy has found that the practical implications of shifts in public attention are quite profound.[18] When the public is highly attentive to an issue, policymakers tend to give high priority to those issues, and policy change becomes more likely.[19] In addition, public attention can shift rapidly when events draw the focus of media and relevant "transmitters" of information related to the event (Kasperson et al. 1988; Kasperson et al. 2001). While trends evident in surveys of public opinion (as described in Section 3) can identify retrospective shifts in public opinion over time, such surveys are available only "after-the-fact," and public opinion measurements on nuclear issues are taken only sporadically and typically with insufficient specificity to meaningfully evaluate the kinds of on-going changes that are of immediate relevance to nuclear facility siting efforts. If the officials leading public engagement programs for ongoing siting initiatives are to quickly and accurately monitor these shifts, and to understand the nature of evolving concerns, real-time mechanisms for tracking and understanding changes in public attention and concerns are necessary.

This section focuses on new methods for tracking and understanding shifts in public attention and opinion. To do so we introduce a third stream of evidence that is sometimes referred to as "*Infoveillance,*" to supplement the quantitative case data and aggregated survey series discussed in prior sections of this report.

4.2. Measuring Public Attention

Public attention is dynamic and unpredictable—it can change with little advance warning. In response, social scientists have devised a number of survey-based metrics for characterizing and estimating public attention. In most instances, these metrics indirectly gage public attention by asking respondents to indicate the issues or problems that they worry about and/or are important to them. If an issue is important, the theory goes, then the public is most likely paying attention. While informative, metrics of this sort suffer from a number of practical limitations, particularly for public officials who are faced with real-time changes that affect public engagement programs. Most notably, the surveys that scholars typically use to develop "importance" metrics are administered on a periodic (i.e., annual or monthly) or intermittent basis. As a result, the cycling of attention that occurs within these broad intervals is often overlooked. Moreover, when changes in public attention are detected, it is difficult to pinpoint when those changes occurred and therefore even more difficult to explain *why* the shift occurred and how it will impact public attention and response in the near future.

To overcome these limitations, researchers are working to advance "real-time" indicators of public attention based on continuous flows of information to supplement the intermittent or periodic data gathered by way of survey research. Infoveillance, which leverages the massive amount of publicly available information that people transmit and/or broadcast via the Internet to determine which issues the public is paying attention to, represents a promising development in this pursuit (Ripberger 2010). For purposes of categorization, most applications of infoveillance can be divided into one of two categories: supply- or demand-based studies (Eysenbach 2009).

Supply-based applications use unsolicited information that members of the public publish on the Internet—via platforms such as Twitter, Facebook, Web pages, blogs, or other social media sites—to capture the ebb and flow of attentiveness to particular issues (like nuclear energy) over time and space. The more people engage in communication, posting, or repeating information about a particular issue, the more likely it is that they are paying attention to it. Thus, the ebbs and flows in issue-specific content indicate variations in public attention.

Demand-based applications, by comparison, use data on the key words and types of information the public is searching for—on Google and other search engines—to track fluctuations in public attention to a given issue. Again, the more information that people seek about a particular issue, the more likely it is that they are paying attention to it. As such, spikes in search traffic can be used as indicators of spikes in public attention.

4.3. Public Attention to Nuclear Issues in the Aftermath of Fukushima

To demonstrate the utility and flexibility of the real-time measures of public attention, the sections that follow use supply- and demand-based indicators to examine fluctuations in public attention to nuclear issues before and after the Tōhoku earthquake and subsequent tsunami that struck Japan on March 11, 2011. Specifically, we track public attention to two different issues—nuclear energy and the management of UNF—between October 2010 and September 2012.

4.3.1. Methods and Data Collection

Consistent with previous research (e.g., Chew and Eysenbach 2010), we created supply-based measures of public attention by systematically archiving publically available "tweets" (user generated content) on Twitter (https://twitter.com/) that contained one or more phrases ("keywords") associated with nuclear energy and/or the management of UNF. Specifically, we measured public attention to nuclear energy by collecting and analyzing the set of 1,084,085 publicly posted tweets that contained the words "nuclear energy" and/or "nuclear power" between October 5, 2010 and September 6, 2012. At the same time, we measured public attention to issues associated with the UNF management by collecting and analyzing the significantly smaller set 180,894 tweets that contained the phrases "Yucca Mountain," "nuclear waste," "radioactive waste," "nuclear dump," "nuclear storage," and "used nuclear fuel" between October 5, 2010 and September 6, 2012. In addition to the text of the tweet itself, we gathered other information about the tweet, including the date/time it was published and the geographic location of the user that published it (when available).

To supplement our supply-based measures, we created demand-based measures of public attention by systematically tracking search queries on Google (e.g., Scheitle 2011; Granka 2010; Reilly, Richey, and Taylor 2012; Ripberger 2010). We used Google Trends (http://www.google.com/trends/) to examine aggregate fluctuations in public searches for the list of issue-specific keywords (as noted above) between October 5, 2010 and September 6, 2012. Note, however, that the search data provided by Google are not given in absolute terms (raw number of searches for a given keyword). Rather, Google Trends scales the data by dividing each number in the series by the highest frequency in that series and then multiplying by 100. The end result of this process is set of values that can range from 0 to 100, with 100 being the highest volume of searches in any given time period.

4.3.2. Analysis of Nuclear Energy/Nuclear Waste Indicators of Attention

Figure 4-1 shows the frequency of tweets containing any of the keywords for nuclear energy from October 2010 through September 2012.

The volume of tweets is indicative of an extraordinary spike in public attention to nuclear energy associated with the event. In the weeks preceding the Fukushima event the average number of tweets on the nuclear terms described above came to 2,740 postings per week (just under 400 per day). The peak traffic, in the week of March 13, 2011, was nearly 155,000 tweets – a fifty-six fold increase over the prior norm. The average weekly traffic volume in the year and a half after the event hovered around 12,800 tweets, holding at a level that is nearly five times the pre-Fukushima volume of attention. This shift in level of attention is statistically significant ($p<0.05$), even when the weekly variation in traffic is taken into account. In short, the Fukushima event triggered an immediate spike of attention to nuclear energy, followed by a sustained and markedly higher new norm in twitter traffic on nuclear issues following the event. The content and characteristics of the posted tweets can be analyzed once the series is obtained ("scraped") and stored in an appropriate database. Table 4-1 shows a sample of tweets from the peak week of twitter traffic (March 13 – March 19, 2011).

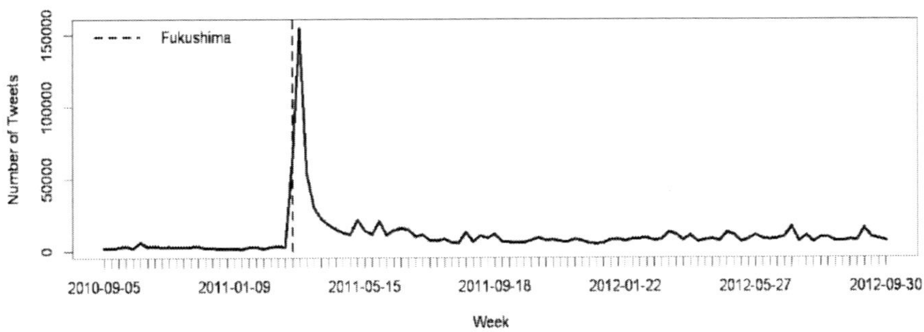

Figure 4-1. Tweets about Nuclear Energy by Week.

Table 4-1. Example "Nuclear Energy" Tweets during Peak Traffic (March 13 – March 19, 2011)

• "Japanese chief cabinet secretary confirms radiation leakage occurred from explosion at #Fukushima nuclear power plant"
• "Watching the news & it's so hard to watch all of this footage of Japan. I feel helpless. Does this nuclear power plant talk scare anyone else?"
• "Tokyo Electric Power Co. says it has lost control of pressure in 2 nuclear power reactors; temps rising"
• "Those who rely on #nuclear power are sitting on a time bomb"

The dataset of tweets also provides opportunities for geographic analysis. Though not all tweets can be geo-coded (individual users must opt-in to having their tweets geo-coded), our analysis suggests that the geo-location of over 8% can be identified. Figure 4-2 shows the breakdown of geocoded tweets concerning "nuclear energy" by the state of residence in the US from which the tweet originated. Note that the breakdown was calculated as follows:

1. First, the number of geo-tagged tweets (those that have a latitude/longitude) about nuclear energy in each state was divided by the total number of tweets in that state.
2. Then we divided the values obtained for each states by the maximum value in that series (Vermont had the highest value) and then multiplied by 100.
3. Vermont, where the Yankee Vermont (YV) power plant is located, had the highest relative frequency of tweets concerning nuclear power. The interaction of concerns about the YV plant with the events at the Fukushima reactors and storage raised the salience of nuclear energy issues to particular heights among Vermont residents. Note that geographic analysis permits identification of tweets at the county or city level, for more focused analyses of frequencies and content.

Figure 4-3 shows the relative frequency of Google searches by week that included any of the "nuclear energy" search terms described above. Recall that Google Trends are reported as the fraction of volume based on the most heavily searched period in the analysis. For these data, as was the case with the Twitter data stream, the peak week was March 13 – 19, of 2011. The most popular search string (the terms typed into Google's search box) utilized in that week was "japan nuclear power."

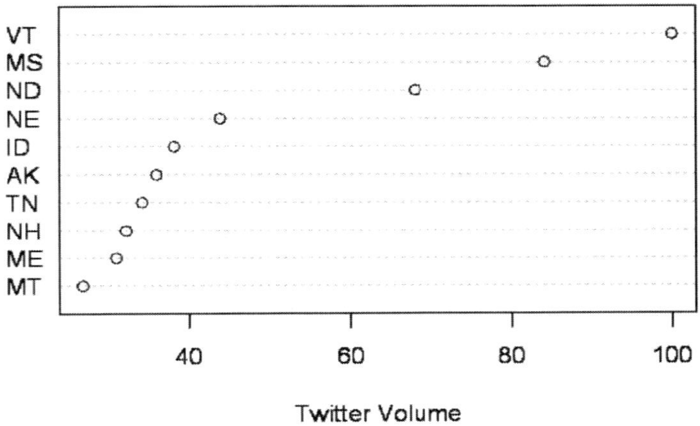

Figure 4-2. Top 10 States by Tweets about Nuclear Energy.

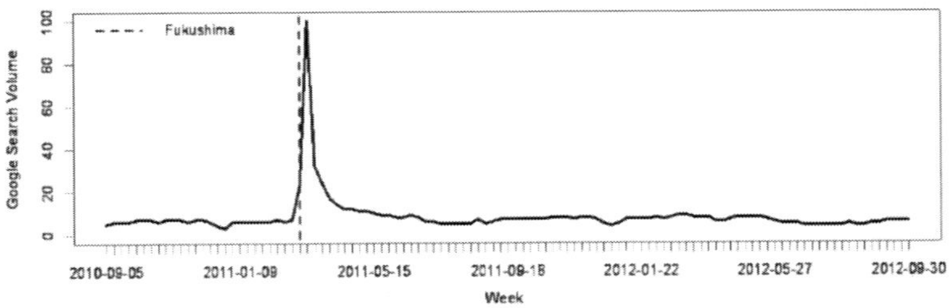

Figure 4-3. Google Searches about Nuclear Energy (by Week).

The spike in searches (using any of the nuclear energy terms listed above) following the Fukushima event was dramatic. Moreover, the long range averages before and after that spike showed a persistent and statistically significant ($p<0.05$) increase in attention; searches averaged a relative search score (compared to the peak volume in the "spike" following the event) of 6.15 is the period prior to the event, but rose to 8.82 over the weeks following the spike. Again, this suggests the development of a "new norm" in attentiveness to nuclear issues following the event.

Like Twitter data, the geographic location of Google searches can be analyzed. Table 4-4 shows the relative Google search volume for nuclear energy by state. As was the case with the Twitter traffic, Vermont was the heaviest source of these searches, followed by the neighboring state of Massachusetts. Next in line were New Mexico, South Carolina and Idaho – all states with federal nuclear facilities.

4.3.3. Analysis of Indicators of Attention to Used Nuclear Fuel Management

Recall that our measures of real-time attention to UNF issues "scraped" Twitter postings including the phrases "Yucca Mountain," "nuclear waste," "radioactive waste," "nuclear dump," "nuclear storage," and "used nuclear fuel." Focusing on the October 5, 2010 to September 6, 2012 time period, our scrape obtained a set of 180,894 tweets. The weekly frequency of tweets over this time period is shown in Figure 4-5.

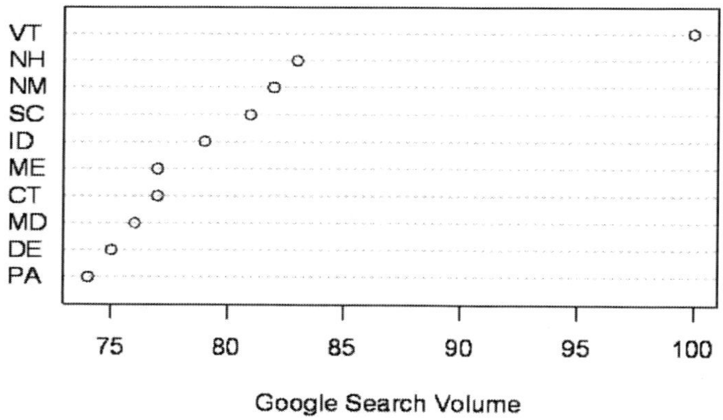

Figure 4-4. Top 10 States by Searches about Nuclear Energy.

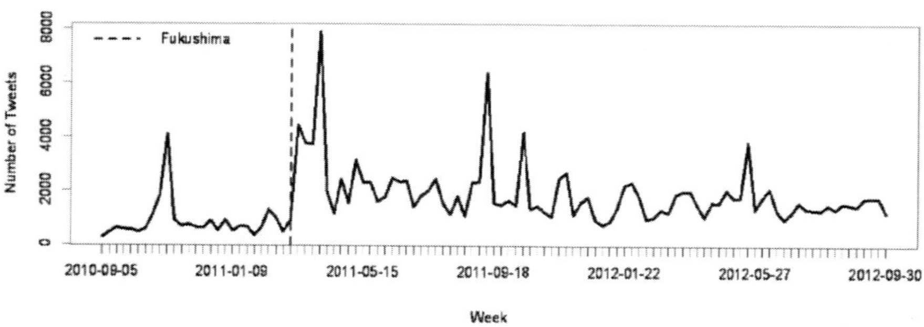

Figure 4-5. Tweets about the Management of Used Nuclear Fuel by Week.

Table 4-2. Example "Used Nuclear Fuel Management" Tweets during Peak Traffic (April 3 – April 9, 2011)

• "7 tons of Radioactive Waste is Flowing into the Ocean Every Hour at Fukushima"
• "Desperate efforts in Japan to stem the tide of radioactive waste spilling into the ocean"
• "I wonder how TEPCO squares "endeavoring sincerely to create a better environment" with dumping radioactive waste in the ocean & atmosphere"
• "Impressed with the tremendous advances in nuclear waste storage. Japan's nuclear officials are simply dumping it all into the ocean"

As is evident in Figure 4-5, there was a surge of twitter traffic following the Fukushima event, peaking at nearly 8000 tweets during the fourth week after the event (April 3-9). As was the case with the nuclear energy/power tweets, the volume of traffic was significantly greater ($p<0.05$) following the event (an average of 1,914 per week) than it was before (when it averaged 849 per week). Table 4-2 shows the content of a sampling of tweets for the period of peak twitter traffic on UNF management.

The regional pattern of tweets about UNF management issues was different than that for nuclear power/energy. As shown in Figure 4-6, the highest relative frequency of tweets on this issue was in Wyoming – a state that had seriously considered hosting a UNF interim storage site in the 1990s until the Governor vetoed the initiative. New Mexico, which hosts the only licensed deep geologic repository in the US, was ranked second in the relative frequency of tweets on this issue.

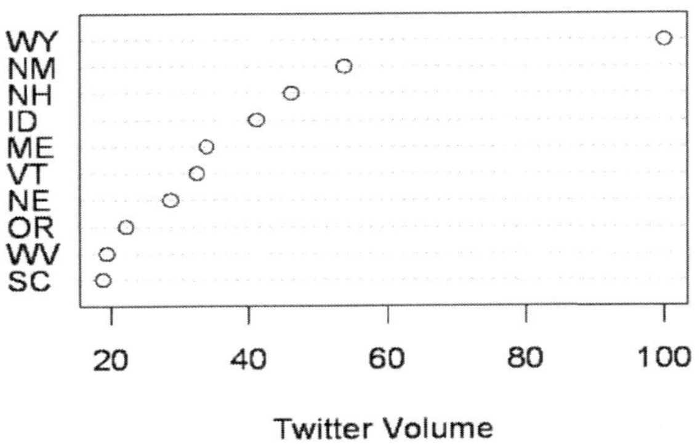

Figure 4-6. Top 10 States by Tweets about the Management of Used Nuclear Fuel.

Shifting now to the demand-side measure, the pattern of Google searches on the UNF management terms is shown in Figure 4-7. Recall that the Google searches are rescaled based on the highest frequency search period (which is scored 100), and all other periods are scored (on a zero to 100 scale) as a percentage of the frequency of searches in that peak period.

As has been the case in all of the series investigated here, the peak period for Google searches on the UNF management followed the Fukushima event. The peak week was March 13-19, 2011, over which the most common single search string (containing one of our search terms) was "nuclear waste Japan." Interestingly, for the full range of nuclear waste searches, there was not a statistically significant change in the overall volume before and after the Fukushima event.

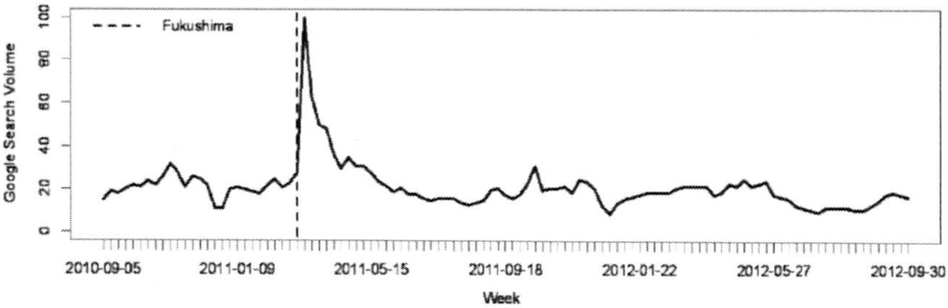

Figure 4-7. Google Searches about the Management of Used Nuclear Fuel by Week.

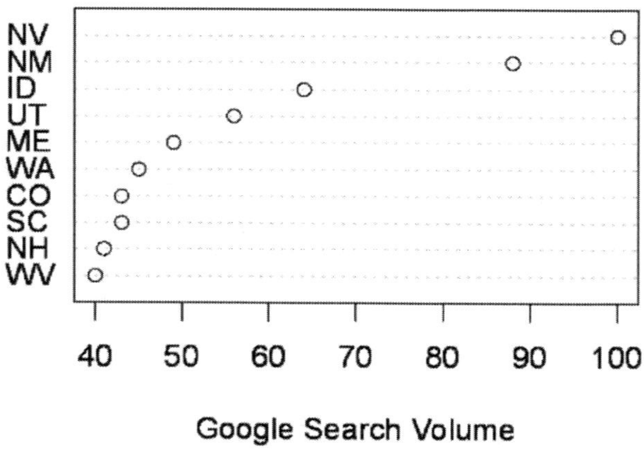

Figure 4-8. Top 10 States by Searches about the Management of Used Nuclear Fuel.

The regional analysis showed the pattern of Google searches to be somewhat different than the pattern of Twitter traffic concerning UNF management. The top states for Google searches were Nevada (host of the now halted Yucca Mountain repository) and New Mexico (site of the WIPP repository).

4.4. Discussion and New Directions

In this section we have sought to describe new ways to track and analyze shifts in public attention on issues like nuclear energy and nuclear waste management, using new kinds of data streams and analysis referred to as *infoveillance*. The patterns of both supply (e.g., Twitter) and demand (e.g., Google searches) provide remarkable, real-time pictures of public attention to nuclear issues before, during and after the Fukushima nuclear crisis event. The pulse of interest and concern following on the heels of the Fukushima event is evident in each of the series, though there are important variations.

The supply streams, based on Twitter traffic, both showed a large pulse of increased traffic, then returned to a new baseline of attention considerably above that which had been evident before the event. The demand streams – in which searches for information that were initiated by users – were less consistent. Searches for nuclear energy information pulsed and then stabilized at a frequency of searches above pre- event searches. Nuclear waste searches also pulsed up following the event, but then returned to pre-event levels. In aggregate, however, these streams indicate a large but temporary spike in attention to nuclear issues following the event, accompanied by a sustained long-term increase in attention to nuclear energy and possibly UNF issues. Accordingly, the Fukushima event has established a new baseline level of attention to nuclear issues.

The geographic distribution of the demand and supply streams, as indicated by the state of origin of the Twitter postings and Google searches, provides the basis for analysis of those regions in which attention rises disproportionately. In nearly all cases, the largest proportionate attention was evident in regions that had experience, of one kind or another, with nuclear energy and/or nuclear waste facility siting. Both of the data streams illustrated here would permit more precise geographic analysis; the Google search data can be identified at the city level, while a fraction of the Twitter data have precise latitude/longitude coordinates for the site of the tweet.

The content of the postings and searches can also be analyzed. Tables 4-1 and 4-2 provide illustrations of the type of content that can be found in the Twitter postings. Note that this content can be analyzed in much greater detail than illustrated here; the types of expression can be coded and tracked, evaluating the kinds of issues that are raised and the emotional valences that are expressed. The weight of that content can be gauged by evaluating the frequency of re-posting (retweets) for particular messages. The networks of communications can also be studied by analyzing the patterns by which users follow and respond to each other. The Google searches, though not messages *per se*, can be analyzed for the combinations and ratios of search terms used (e.g., negative and positive terms used in combination with the search terms).

In short, there are rich possibilities for analyzing the ways in which attention is distributed, the specific issues that are of most concern, the emotive content of those concerns, and the patterns of interaction about those concerns. These types of analyses can be implemented in real time as issues evolve, providing program managers with real-time tracking of public concerns that can inform more responsive and effective communication and engagement programs. More generally, infoveillance and the associated analyses can provide feedback on the nature and breadth of public concerns in ways that can assist in the design and implementation of consent-based facility siting programs.

5. National Survey of Energy and Environmental Issues: 2012

5.1. Survey Design and Implementation

The preceding sections of this report have focused on three distinct lines of evidence that can be used to inform the design and implementation of consent-based siting programs for UNF storage and disposal facilities. As described in Section 2, the global database of siting cases enables analysis of the characteristics of historical siting efforts that systematically influenced the outcome of that effort. Section 3 analyzed aggregated measures of US public opinion taken from hundreds of opinion polls over the period from 1973 through the present, focusing on the manner in which public "mood" regarding nuclear issues has been influenced by major nuclear crisis events, as well as identifying the underlying trends operating independently of particular events.

Section 4 turned to a set of "real time" measure of shifts in public attention to nuclear issues. Analysis of each of these distinct streams of evidence addresses a different aspect of the social context in which efforts to site nuclear facilities have taken place. None of them, however, can directly address public perceptions and preferences for *future* siting initiatives employing the kinds of consent-based approaches advised by the BRC. To look prospectively at how the public would understand and respond to this alternative approach, a specially designed and implemented survey dedicated to UNF management issues is required.

This section summarizes findings from an Internet survey of 2,017 respondents conducted 1–5 June, 2012, measuring US public views on selected energy security issues that help shape the context for policy debates about the energy future. Key issues include assessments of energy policies, confidence in future energy sufficiency, preferences for energy sources, and the implications of beliefs about climate change for energy preferences. Special emphasis was given to public views on nuclear energy and the disposition of UNF.[20]

Probabilistic sampling of the US general public and lengthy interviews historically done by landline phones are no longer feasible due to rapidly evolving telecommunications patterns and the impracticality of conducting extensive surveys on complex policy domains using wireless phones. Our sample of respondents was randomly drawn from the SurveySpot Internet panel of more than a million members of the US general public maintained by Survey Sampling, International. Participants received a small remuneration ($5) for their participation. Though the results do not represent a random sample of the entire US population, the demographic attributes of survey respondents are set to mirror the demographic composition of the American general public as determined by the most recent US Census estimate. Previous annual surveys in this series since 2006 show a high degree of continuity in response patterns and relational analyses. A complete list of the survey questions and the distributions of responses from the 2,017 respondents are included in Appendix C.

5.2. Survey Findings

5.2.1. Perspectives on Energy and Climate Change

To help place energy issues within a larger national policy context, we traditionally ask participants to indicate their level of concern about five broad policy issue areas listed in random order: national security, the economy, healthcare, energy, and the environment. On average, each issue area was ranked high in absolute measures of concern (means well above midscale). The relative hierarchy of concerns is shown in Figure 5-1, with the availability and cost of energy being ranked lower than the state of the US economy/jobs and the cost and delivery of healthcare ($p<0.01$), but notably higher than national security (including terrorism) and the effects of human activities on the environment ($p <0.0001$ each pairing).

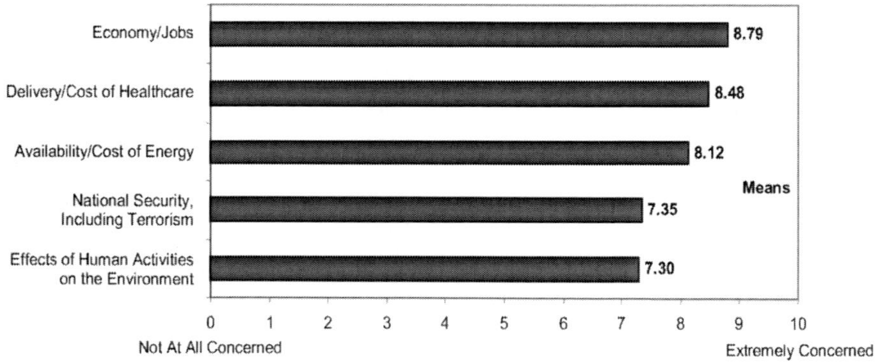

Figure 5-1. Relative Issue Concerns.

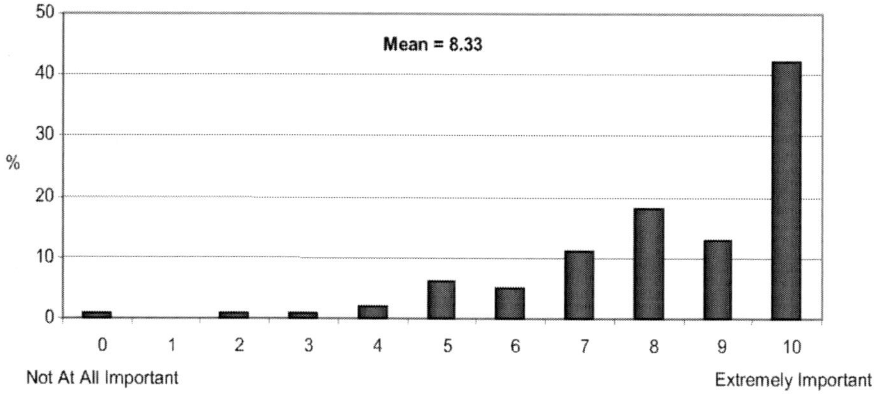

Figure 5-2. Perceived Importance of Reducing US Energy Dependence.

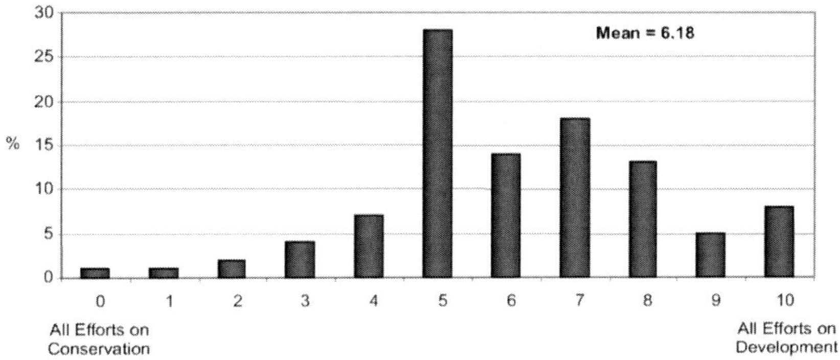

Figure 5-3. Preferences for Balancing Energy Development and Conservation.

Narrowing the policy focus to energy, we traditionally ask a series of questions on satisfaction with current energy policies, confidence in the adequacy of future energy supplies, the importance of reducing US dependence on foreign sources of energy, and preferences for balancing energy development and conservation. In 2012, about 48% of respondents indicated dissatisfaction with current energy policies; 30% reported varying levels of satisfaction, and 23% were unsure. Notwithstanding general unease with current energy policies, a majority of participants (57%) expressed confidence in having adequate sources of energy to meet US needs during the next 20 years. A clear consensus was expressed about two key elements of the energy future—US energy independence and balancing energy development with conservation. As shown in Figure 5-2, when asked to express the importance of reducing US dependence on foreign sources of energy, the modal response was the highest point on the zero-to-ten scale, and fully 89% of respondents rated reducing energy dependence above midscale.

When asked to express their preference for balancing energy development and conservation, Figure 5-3 shows that 15% of the survey respondents preferred emphasizing conservation, 28% wanted an equally balanced strategy, and a majority of 58% preferred to emphasize energy development.

Preference for energy development over energy conservation increased systematically with age and income and was significantly higher among men, political conservatives, and those residing in rural settings.

To gage overall perceptions of the risks associated with energy sources, we traditionally ask participants to rate perceived risks of fossil, nuclear, and renewable sources of energy. Figure 5-4 compares perceived risks for the three energy sources across a continuous scale where zero represents no risk and ten represents extreme risk.

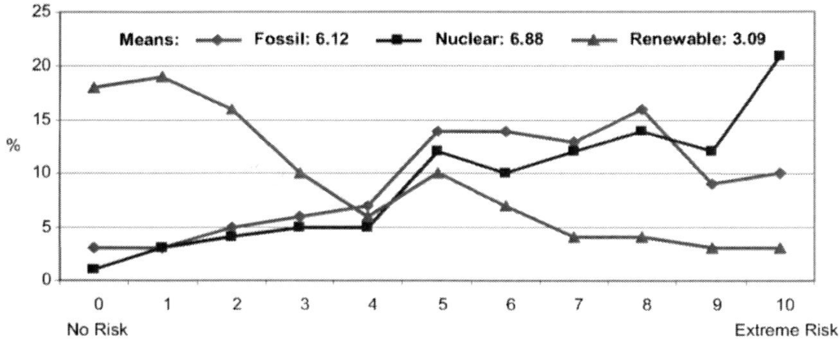

Figure 5-4. Comparing Energy Risk Perceptions.

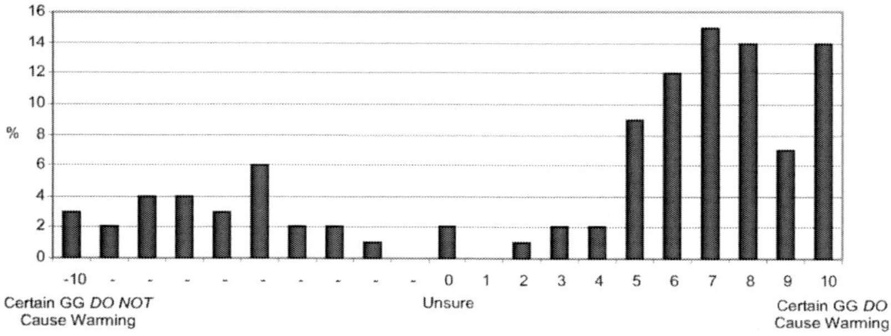

Figure 5-5. Beliefs on Climate Change.

Nuclear energy was perceived to pose the highest mean risk, followed by fossil fuels. Renewable sources were judged to pose the least risk. Differences in means are statistically significant ($p < .0001$) for each paring.

When we informed respondents of the current proportions of total US energy and of electricity generation deriving from fossil, nuclear, and renewable sources and asked them to identify how they would prefer the mix of sources for energy and electricity to evolve over the next 20 years, most respondents favored reducing proportions of energy deriving from fossil fuels, modestly increasing the proportion of total energy provided by nuclear generation, and substantially enlarging overall production from renewable energy sources. But when pressed further on the prospects for living with fossil fuels, only 22% opposed further exploration and development of US deposits of oil and gas; opinions were almost evenly divided on the emerging issue of using "fracking" (hydrological fracturing) techniques to extract hard-to-reach deposits of oil and natural gas; and a majority of respondents (53%) thought the proposed Keystone XL Pipeline should be constructed in the US (about 23% were unsure, and 24% opposed the pipeline). So while our respondents preferred to reduce dependence on fossil fuels over the long-term, most supported continued development of US oil and gas reserves in the short-term and using them to help reduce American dependence on foreign energy sources.

To investigate how beliefs on global climate change (GCC) relate to energy preferences, we asked participants whether they believed greenhouse gases, such as those from the combustion of fossil fuels, are causing global temperatures to rise, and how certain they were of those beliefs. Figure 5-5 shows responses arranged along a scale from those completely certain that greenhouse gases do not cause global warming to those completely certain that greenhouse gases do cause warming.

About 78% of respondents believed greenhouse gases are causing average global temperatures to rise, and they were substantially more certain of their beliefs than were those who did not believe a causal relationship exists between greenhouse gases and warming. Respondents 50 years of age and above, men, and political conservatives were significantly more doubtful of the cause and effect relationship, while younger respondents, women, and political liberals were most sure that greenhouse gases cause global warming. To relate these beliefs about global climate change to the energy future, we looked at the broad energy preferences previously described as a function of certainty about the causes of global climate change. Participants who were confident that greenhouse gas emissions contribute to global warming expressed statistically significantly lower preferences for fossil fuels as future sources of energy and significantly higher preferences for renewable sources. But they also preferred to reduce the proportions of future energy provided by nuclear generation, even though it produces negligible amounts of greenhouse gas emissions. Some of this desire can be attributed to traditional opposition to nuclear energy by environmentalists, but it also partially is a function of members of the public not understanding that nuclear generation does not emit appreciable greenhouse gases. When asked to agree or disagree with the assertion that nuclear power plants produce significant amounts of greenhouse gases, only about 36% disagreed, while 37% agreed with the false assertion, and 28% were unsure. Clearly, associations of nuclear energy with global climate change are confused by a lack of factual understanding that nuclear generation does not produce significant amounts of greenhouse gases.

5.2.2. Nuclear Energy Beliefs

To bring the longer-term trends in public views on nuclear energy discussed in Section 3 into sharper focus for 2012, we asked participants to rate four categories of potential risks associated with nuclear energy, and four types of potential benefits. Using a scale from zero (no risk) to ten (extreme risk) respondents rated perceived risks associated with each of the following:

- An event at a US nuclear power plant within the next 20 years that results in the release of large amounts of radioactivity
- An event during the transportation or storage of UNF from nuclear power plants in the US within the next 20 years that results in the release of large amounts of radioactivity
- A terrorist attack at a US nuclear power plant within the next 20 years that results in the release of large amounts of radioactivity
- The diversion of nuclear fuel from a nuclear power plant in the US within the next 20 years for the purpose of building a nuclear weapon

To assess potential benefits, we asked participants to rate the following four items using a scale from zero (not at all beneficial) to ten (extremely beneficial):

- Fewer overall greenhouse gas emissions because nuclear energy production does not create greenhouse gases
- Reliable power because nuclear energy generates large amounts of electricity and is not affected by weather conditions, such as low rainfall or no wind
- Greater US energy independence because nuclear energy production does not require oil or gas from foreign sources
- Reduced environmental damage because of less need for mining coal or extracting oil and gas

Table 5-1 summarizes mean responses ranked from highest to lowest.

Note that a potential terrorist attack was judged as the highest risk, and nuclear energy's contribution to US energy independence was rated as its greatest benefit. Mean nuclear energy risks were statistically significantly higher among respondents under 50 years of age, women, those without a college degree, racial/ethnic minorities, participants with lower annual household incomes, and individuals who were politically more liberal. Conversely, mean benefits of nuclear energy were significantly greater among men, those with a college degree, participants with higher annual household incomes, and those who were politically more conservative.

To require participants to integrate and comparatively weigh the relative risks and benefits of nuclear energy, we asked the following question, and chart results in Figure 5-6.

Using a scale from one to seven, where one means the risks of nuclear energy far outweigh its benefits, four means the risks and benefits are equally balanced, and seven means the benefits of nuclear energy far outweigh its risks, how do you rate the overall balance of the risks and benefits of nuclear energy in the US?

Table 5-1. Mean Nuclear Energy Risk and Benefit Assessments

Perceived Risks: 0 = No Risk -10 = Extreme Risk		**Perceived Benefits 0 = Not At All Beneficial— 10 = Extremely Beneficial**	
Terrorist Attack	6.57	Energy Independence	7.04
Operational Accident	6.42	Reliable Power	6.89
Transportation Accident	6.19	Less Mining / Extraction	6.75
Diversion to Nuclear Weapons	5.70	No GG Emissions	6.74

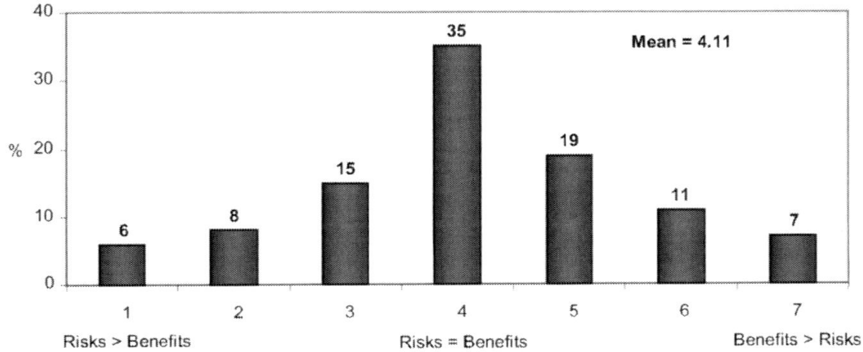

Figure 5-6. Balancing Nuclear Energy Risks and Benefits.

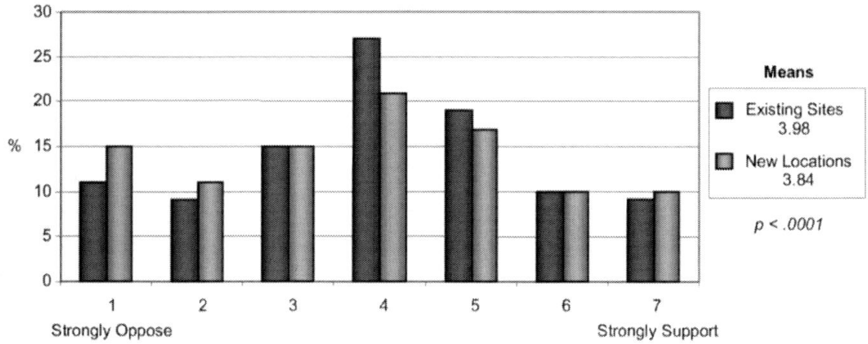

Figure 5-7. Support for Constructing Additional Nuclear Power Reactors.

About 37% of respondents considered nuclear energy benefits to outweigh risks, while a similar number (35%) thought the risks and benefits of nuclear energy were equally offsetting, and the remaining 29% of participants considered the risks of nuclear energy to outweigh its benefits.

Next, we asked participants to indicate their level of support for (a) constructing additional nuclear reactors at the sites of existing nuclear power plants in the US, and (b) constructing additional nuclear power plants at new locations in the US. Figure 5-7 compares responses.

While mean support was significantly higher for constructing additional reactors at existing sites as opposed to constructing new nuclear power plants at additional sites, notice that opinion was quite divided on both questions, with opposition to additional reactors at existing sites (34%) being similar to support (38%). Also notice that 27% of respondents were undecided. The proportions of opposition and support for building additional nuclear power plants at new locations also were quite divided, with 41% opposing, 37% supporting, and 21% undecided.

These widely divided opinions on additional nuclear generation raise questions about the underlying belief structures that shape public views on nuclear energy. By averaging responses to our question on support for additional nuclear generation at existing locations and responses to our question on support for constructing new nuclear power plants, we created an index representing support for additional nuclear generation and used that as the dependent variable in a causal model to further investigate relationships among perceptions and beliefs about nuclear energy.

To construct a causal model explaining support for new nuclear generation, we drew on extensive literature about hierarchically structured belief systems (Fiske and Taylor 1992; Herron and Jenkins- Smith 2002, 2006; Hurwitz and Peffley 1987; Hurwitz, Peffley and Seligson 1993; Jenkins-Smith and Herron 2002a, 2002b, 2005; Jenkins-Smith, Mitchell, and Herron 2004; Peffley and Hurwitz 1985; and Sabatier and Jenkins-Smith 1993, 1999). Our model includes the following four hierarchical levels or stages.

Demographic predispositions pre-exist beliefs and may condition their formation. Our model includes the following five measures of demographic attributes.

(a) Respondent age is represented by a continuous variable from 18 to 94.
(b) Education is represented by an indicator variable[21] in which a value of one represents a four year college degree or higher level of education, and zero represents less than that level of education.
(c) Gender is represented by an indicator variable in which a value of one represents men, and a value of zero represents women.
(d) Household income for the year 2011 is represented by a continuous variable having 21 increments of $10,000 from a minimum of <$10,000 to a maximum of >$200,000.
(e) Race/ethnicity is represented by an indicator variable in which a value of one represents racial/ethnic minority status, including American Indians, African Americans, and Hispanics, and a value of zero represents all other races and ethnicities.

Core beliefs are the most general and abstract, consisting of fundamental underlying normative dispositions that transcend specific policy issue domains. Our model includes the following five measures of core beliefs:

(a) Government trust is represented by a generalized measure of how much of the time the respondent expects the federal government "to do what is right for the American people," expressed on a continuous scale from zero (none of the time) to ten (all of the time).
(b) The Natural environment index is a composite measure of beliefs about the fragility/resilience of nature and the state of the environment expressed on a continuous scale from zero (nature is robust and the environment is not at all threatened) to ten (nature is fragile and the environment is on the brink of disaster).

(c) Individualism is measured on a continuous scale of agreement from zero (not at all) to ten (completely) with the following characterization of political culture:[22]

> Groups are not all that important to me. I prefer to make my own way in life without having to follow other peoples' rules. Rewards in life should be based on initiative, skill, and hard work, even if that results in inequality. I respect people based on what they do, not the positions or titles they hold. I like relationships that are based on negotiated "give and take," rather than on status. Everyone benefits when individuals are allowed to compete.

(d) Hierarchism is measured on a continuous scale of agreement from zero (not at all) to ten (completely) with the following characterization of political culture:

> I am more comfortable when I know who is, and who is not, a part of my group, and loyalty to the group is important to me. I prefer to know who is in charge and to have clear rules and procedures; those who are in charge should punish those who break the rules. I like to have my responsibilities clearly defined, and I believe people should be rewarded based on the position they hold and their competence. Most of the time, I trust those with authority and expertise to do what is right for society.

(e) Egalitarianism is measured on a continuous scale of agreement from zero (not at all) to ten (completely) with the following characterization of political culture:

> Much of society today is unfair and corrupt, and my most important contributions are made as a member of a group that promotes justice and equality. Within my group, everyone should play an equal role without differences in rank or authority. It is easy to lose track of what is important, so I have to keep a close eye on the actions of my group. It is not enough to provide equal opportunities; we also have to try to make outcomes more equal.

Domain beliefs reflect fundamental orientations and strategies that apply across a specific policy issue domain such as nuclear energy. Our model includes the following four measures of domain beliefs.

(a) Trust in federal information is a more specific composite (average) measure of trust in information provided by science and engineering experts from the Nuclear Regulatory Commission, the Environmental Protection Agency (EPA), US national laboratories, and the DOE. It is expressed on a continuous scale from zero (no trust) to ten (complete trust):

(b) The nuclear energy risk index is a combined average measure of four types of potential risks associated with nuclear energy described above and summarized in Table 5-1. It is expressed on a continuous scale from zero (no risk) to ten (extreme risk).

(c) The nuclear energy benefit index is a combined average measure of four types of potential benefits deriving from nuclear energy also described above and summarized in Table 5-1. It is expressed on a continuous scale from zero (not at all beneficial) to ten (extremely beneficial).

(d) The GCC certainty index represents the degree of certainty respondents expressed about whether greenhouse gas emissions are or are not causing average global temperatures to rise. Described above and displayed in Figure 5-5, it is represented on a continuous scale from –10 (completely certain that greenhouse gases are not causing warming) to +10 (completely certain that greenhouse gases are causing warming).

Policy preferences are beliefs about issue priorities and preferred policy choices within a given policy domain. For modeling purposes, responses to our questions on support for additional nuclear generation at existing locations and support for constructing additional nuclear power plants at new locations were averaged to form an index expressed on a continuous scale from one (strongly oppose) to seven (strongly support). Preferences represented by the additional nuclear generation index provided the dependent variable being explained by the causal model shown in Figure 5-8.

The relationships depicted in Figure 5-8 were calculated sequentially using multivariate regressions to identify standardized coefficients among those relationships that were statistically significant at the 95% confidence level ($p < .05$). Only statistically significant relationships are shown in the model. In the first stage, we used the five demographic measures described above to explain variation in each of the five core belief measures. In the second stage, we combined the five demographic predispositions with the five core belief measures to explain each of four domain beliefs. In the final stage, we completed the model by combing demographic attributes, core beliefs, and domain beliefs as independent variables in a multivariate regression to explain support for additional nuclear generation at existing and new locations. This iterative process shows which independent variables act through intermediate variables (beliefs) and which act directly on the final dependent variable. It also reveals the influence of each independent variable when all other independent variables are held constant.

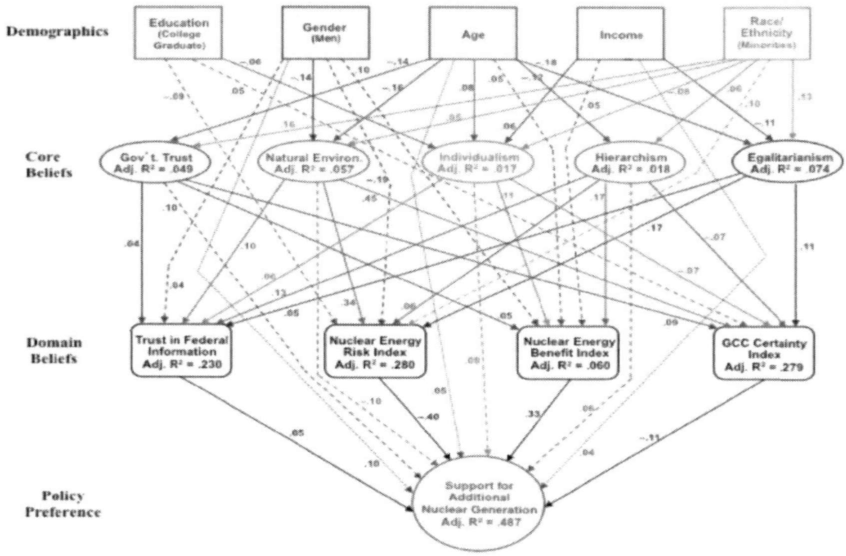

Figure 5-8. Causal Model of Support for Additional Nuclear Generation.

The direction and size of the standardized regression coefficients (shown as numerical values adjacent to causative lines in the model) are interpreted as follows: a change of one standard deviation (*SD*) in the independent variable resulted in the fractional change of a *SD* in the dependent variable represented by the standardized coefficient. For example, a standardized coefficient of 0.25 means that a change of one *SD* in the independent variable caused a positive change of 0.25 *SD* in the dependent variable. In the case of an indicator variable, the coefficient represents the relationship between the independent attribute coded as a value of one and the dependent variable. Because the coefficients are all standardized, they can be compared. Explanatory powers are expressed as adjusted R^2 values. Solid lines represent first order relationships between independent and dependent variables in adjacent echelons of the model; dashed or dotted lines depict relationships extending beyond the adjacent echelon.

The model shows that respondent beliefs about risks and benefits of nuclear energy were the most powerful direct predictors of support for additional nuclear generation. As assessments of nuclear energy risks increased one *SD*, support for additional nuclear generation declined by –0.40 *SD*, and as assessments of nuclear energy benefits increased one *SD*, support for additional nuclear generation increased 0.33 *SD*. In addition to perceptions of nuclear energy benefits, other direct positive influences for additional nuclear generation included general trust in the federal government as well as trust in responsible federal agencies to provide information about nuclear energy; individualism; hierarchism; age; income; and greater support among men than women. In addition to nuclear energy risk perceptions, other direct negative influences on support for additional nuclear generation included concerns about the fragility of nature and the state of the environment; increasing certainty that greenhouse gas emissions contribute to global warming (notwithstanding the fact that nuclear generation produces negligible emissions of greenhouse gases), and lower support among women. Overall, our causal model explained about half (0.49%) of the variation in support for additional nuclear generation measured in this survey.

5.2.3. Implications of Fukushima on Perceptions of Nuclear Energy

As discussed above in Sections 3 and 4, the historical effects of nuclear events on public beliefs about nuclear safety and preferences for reliance on nuclear energy have been sizable, sustained and negative. In order to measure the effects of the Fukushima nuclear event on current preferences for nuclear energy, we asked the survey respondents to the 2012 survey the following question:

> As you may recall, a severe earthquake occurred on March 11, 2011, in the Pacific Ocean near Japan, creating large tidal waves that destroyed some Japanese coastal cities. Also damaged was the Fukushima nuclear power plant, which released radioactivity into the atmosphere and nearby portions of the sea. The earthquake and tidal wave killed thousands of people; the release of radiation at Fukushima is not known to have produced any deaths, but could contribute to future illnesses. We would like to know how the Japanese experience has influenced your confidence in US nuclear power.[23]
>
> On a scale from minus ten to plus ten, where minus ten means the Japanese experience has strongly reduced your support for U.S. nuclear power production, zero means the Japanese experience has had no effect on your support, and plus ten means the Japanese experience has strongly

increased your support, how have recent events in Japan influenced your support for nuclear power production in the United States?

The responses are shown in Figure 5-9.

As indicated in Figure 5-9, while the modal response was "no effect," the average response was that the effect of the Fukushima event was to reduce support for reliance on nuclear power in the US (a drop of -1.5 points, on the 20-point scale). While not large, the direction of this effect is consistent with the historical data shown in Section 3 and 4.

5.2.4. Understanding Public Preferences for Used Nuclear Fuel Management

Measuring public views on technically complex policy issues, such as managing UNF, is most effectively accomplished using a phased approach (Jenkins-Smith et al. 2012a). In the first phase, we measure understandings of selected key issues and current policies to establish a baseline of existing awareness and information. In the second phase, we provide basic factual information and carefully balanced arguments about selected aspects of managing UNF. This stage accomplishes two important prerequisites to policy evaluation: it establishes a foundation of shared accurate factual information, and it exposes respondents to key points in alternative strategies and policy choices. Only then can preferences for policy alternatives be placed in relative context for the third stage of inquiry that asks ordinary Americans to express preferences for policy choices in managing UNF. We began the first stage with two questions that asked participants how UNF in the US currently is being managed, and whether or not UNF currently is being stored at any sites within their state of residence. We presented participants with four randomly ordered response options for how UNF currently is being managed: (a) stored in cooling pools or special containers at nuclear power plants throughout the US; (b) shipped to Nevada and stored in a facility deep underground; (c) chemically reprocessed and reused; and (d) shipped to regional storage sites. About four in ten respondents (39%) knew that most UNF today is being stored on-site in cooling pools or special containers at nuclear power plants or closed nuclear facilities. Only 14% were aware that UNF was being stored temporarily using such methods in their state of residence.

In Stage 2, we provided the following factual information to establish a shared minimum level of knowledge about managing UNF.[24]

Used nuclear fuel is highly radioactive and must be safeguarded for thousands of years or chemically reprocessed. If it is reprocessed, the uranium can be separated from the waste and reused to make new fuel rods for generating electricity, but the remaining elements are highly radioactive for a very long time and must be safeguarded and isolated from the environment for thousands of years.

In 2010 the government halted construction of a deep underground facility inside Yucca Mountain in Nevada that had been intended for permanent storage and disposal of used nuclear fuel, and very little used nuclear fuel is being reprocessed in the US.

After two years of study, in January 2012 the President's Blue Ribbon Commission on America's Nuclear Future recommended quick efforts to build one or more underground nuclear repositories for permanent storage and disposal of used nuclear fuel. Another recommendation was to build one or more interim sites for managing and temporarily storing used nuclear fuel. The Commission also recommended making preparations for transporting nuclear materials to those storage and disposal facilities (Blue Ribbon Commission on America's Nuclear Future 2012).

Currently, US used nuclear fuel is being temporarily stored at over 100 sites in 39 states. Most of it is stored at nuclear power plants where it is placed in specialized concrete cooling pools. In some cases, the used fuel is transferred to specialized concrete casks stored above ground near the nuclear power plant. At each site, the cooling pools and storage casks are protected at all times by security forces. This poses a problem at nine sites where nuclear power plants have been shut down but the used nuclear fuel stored there continues to require expensive security measures that otherwise would not be needed. Some people think that temporarily storing used nuclear fuel at existing sites is an acceptable solution for the foreseeable future, while others think such practices are risky and other options need to be adopted.

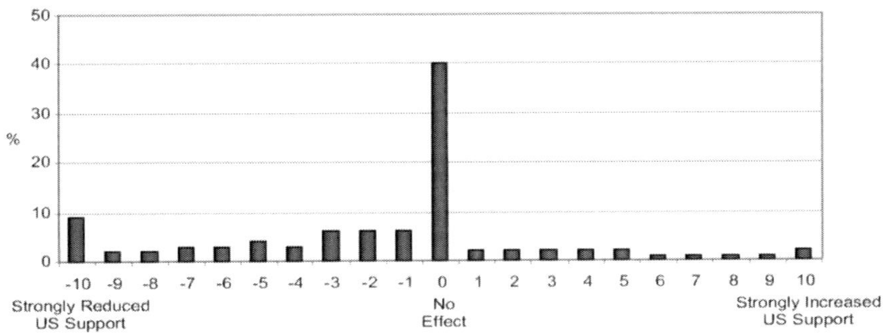

Figure 5-9. Implication of Fukushima Event on US Nuclear Preference.

We then provided in Stage 3 the following randomly ordered condensed counterarguments for and against continuing current practices for on-site storage of UNF.

Opponents argue that some nuclear power plants where used nuclear fuel is stored are near rivers and oceans where flooding is possible. And some of these nuclear power plants are near large population centers. On rare occasions used fuel has leaked radiation into the cooling pools. Moreover, the cooling pools and containers are located at ground level, and therefore might be vulnerable to terrorists. They note that these storage practices do not provide a permanent solution for managing used nuclear fuel.

Supporters argue that transporting used nuclear fuel by train or truck to consolidated storage facilities is risky, that storing used nuclear fuel at nuclear power plants is less expensive than consolidated storage, and that it buys time for finding future solutions. Moreover, storage at nuclear power plants has not caused any accidents in the United States that have exposed the public to radiation.

Having provided a minimum shared base of factual information about current UNF practices and presented arguments for and against the current policy, we then asked respondents to express their support for or opposition to current temporary on-site storage of UNF using a scale from one (strongly oppose) to seven (strongly support). We chart responses in Figure 5-10.

A plurality of respondents (46%) opposed current UNF disposition policies, about 24% supported them, and the remaining 30% were undecided. These results suggest that when apprised of current UNF disposition practices, there is substantial policy space for change, with only about one in four of our participants favoring current policies.

5.2.5. Policy Implications of Repository Concepts, Designs, and Incentives

Next we randomly assigned respondents to one of two tracks in order to better explore new facility design preferences. Half of our participants were asked to consider constructing two underground mine-like permanent repositories for UNF, and half were asked to consider constructing one or more above-ground interim storage facilities for UNF. Each is described below.

Repository Track (50% of respondents): For the next few questions, assume that construction of two underground mine-like repositories is being considered for long-term storage and disposal of used nuclear fuel. One would be in the eastern US, and the other in the west. Each of these sites would include secure surface buildings and a mine deep underground where radioactive materials could be isolated from people and the environment and could be designed to allow retrieval or to permanently seal away the materials. The facilities and the mines would be designed to meet all technical and safety requirement set by the US Nuclear Regulatory Commission, the US Environmental Protection Agency, and state regulatory agencies.

Interim Storage Track (50% of respondents): For the next few questions, assume that construction of one or more interim above-ground storage facilities is being considered where used nuclear fuel could be stored safely for up to a hundred years. Each of these sites would include secure surface facilities where used nuclear fuel could be consolidated and stored, and where the radioactive materials could cool and be prepared and packaged for later shipment to a permanent repository. These interim storage facilities would be designed to meet all technical and safety requirements set by the US Nuclear Regulatory Commission, the US Environmental Protection Agency, and state regulatory agencies.

When members of each track were asked to express their support for or opposition to the described facilities, they responded as shown in Figure 5-11.

Those respondents who received the option for two underground permanent repositories indicated a statistically significantly higher level of support (4.65 on the one-to-seven scale), on average, than did those who received the option for one or more above-ground interim storage facilities (4.29). Both were rated higher than support for continuing temporary on-site storage practices (3.51), but the difference in support shown in Figure 5-11 may indicate that the rational or logic of permanent repositories is more intuitive to members of the public than is the concept of interim storage facilities that requires moving UNF twice. The BRC recommended both permanent and interim storage facilities, with a strong argument for their independent utility and for their synergistic logic, but making the case for public support of both options may require specialized explanations of why both types of facilities are warranted.

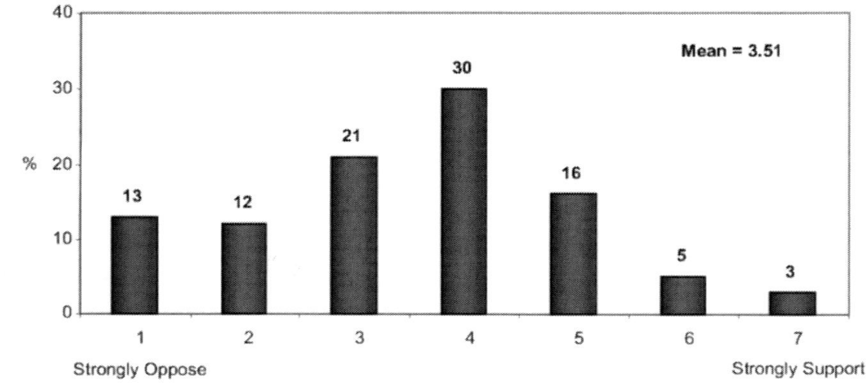

Figure 5-10. Support for Current UNF Disposition on Site.

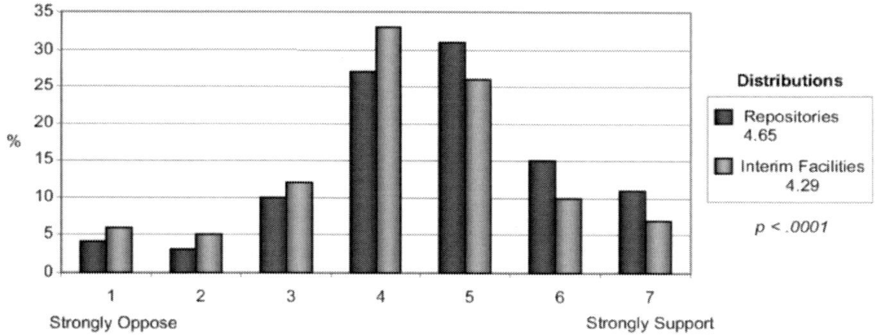

Figure 5-11. Support for Permanent Repositories and Interim Storage Facilities.

By continuing the two tracks—permanent repositories and interim storage facilities—and using the above baseline support metrics, we explored potential implications of varying three facility attributes: (a) co-locating a national research laboratory at the storage sites to develop and design ways to more safely manage, transport, and dispose of UNF; (b) co-locating reprocessing capabilities with storage facilities to permit the reuse of reprocessed uranium for regenerating electricity;[25] and (c) providing varying amounts of compensatory incentives to communities and states that host repositories or interim storage sites.[26] Compensation amounts were randomly assigned as 10, 25, 100, and 300 million dollars per year for each of the two types of facilities. Table 5-2 shows how baseline support for two permanent repositories varied when changes in facility design and incentives were considered, and Table 5-3 shows variations in support for interim storage facilities when the same design attributes and incentives were evaluated.

Table 5-2. Implications of Repository Design and Incentives

Design Attributes	% Oppose	% Unsure	% Support	Means (1–7)	Change from Base Mean
Two mine-like repositories (base case)	**17**	**26**	**57**	**4.65**	**N/A**
With research lab	9	18	73	5.21	+ 12.0 %
With reprocessing	14	24	62	4.84	+ 4.1 %
With compensation	16	23	61	4.81	+ 3.4 %

Table 5-3. Implications of Interim Storage Design and Incentives

Design Attributes	% Oppose	% Unsure	% Support	Means (1–7)	Change from Base Mean
One or more interim storage facilities (base case)	23	33	43	4.29	N/A
With research lab	13	23	64	5.00	+ 16.6 %
With reprocessing	17	25	59	4.72	+ 10.0 %
With compensation	20	22	58	4.68	+ 9.1 %

Several points are evident in Tables 5-2 and 5-3. Regarding physical design characteristics, the addition of a collocated research lab to further scientific understandings of how best to manage, transport, and dispose of UNF, and the addition of reprocessing facilities to reuse uranium from UNF, increased support appreciably for both types of UNF facilities. Second, compensation to host communities and states exerted a modest but positive effect on support.[27] Finally, while interim storage facilities received significantly less baseline support than did permanent repositories, the marginal gains with enhanced design and compensation were larger for the interim sites than for the repositories.

These data regarding UNF facility concepts, designs, and incentives suggest that our respondents preferred permanent repositories and/or interim storage facilities over current temporary on-site storage practices, but the rational and benefits of permanent repositories seemed more persuasive than those for interim storage facilities that require UNF to be transported twice. Support for either permanent repositories or interim storage facilities varied with physical design characteristics and financial compensation (beyond associated jobs and indirect economic benefits), and these data suggest that tailoring facility concepts, designs, and incentive structures potentially may influence public support for or opposition to siting options.

5.2.6. Implications of UNF Facility Siting Process

The survey data provide a number of important implications for how the policy processes for siting UNF storage and disposition facilities may affect prospects for public support or opposition. The BRC recommended pursuing a consent-based approach to policy design, but recommended that specifics of the siting process not be prescribed in advance, and that these be reserved for negotiations with potential host communities and states. The BRC also emphasized the need to obtain consent from prospective hosts, but again, recommended that the characteristics of process for obtaining consent be included in the negotiations. Accordingly, our initial inquiries into public receptivity to potential siting processes necessarily were generalized, focusing on four key issues discussed below: (a) conceptual siting process design preferences; (b) institutional trust in information about UNF management and expectations of risk bias among organizations likely to be involved in design and evaluation of specific siting processes; (c) anticipating issues involved in the process of "consent," and (d) proximity concerns.

***A. Facility Design Concepts*:** It is cognitively challenging for survey respondents to evaluate conceptual approaches to policy designs when important aspects of those designs are reserved for negotiations. We began by contrasting two generalized approaches: (a) "top-down" deductive policy processes that begin with federal entities searching for technically feasible sites and then directing responsible state and local communities to participate in siting negotiations, versus (b) "bottom-up" inductive policy processes in which potential host communities and their states nominate themselves for technical consideration of site feasibility and negotiations over facility design and function. These kinds of general descriptions of policy designs are too vague to elicit firm opinions from most respondents, but they can help suggest directional preferences.

Continuing to employ the split design described above in which half of our respondents received questions relating to the siting of two permanent repositories, and the other half received questions about siting one or more interim storage facilities, we asked each subgroup to consider the following two very generalized descriptions of competing policy designs. Each of the concepts was preceded by stipulating that government regulators would be required to evaluate prospective sites to ensure they could safely contain nuclear materials for the intended period of time in which UNF would be stored (thousands of years for permanent repositories; up to a hundred years for interim storage facilities). The two randomly ordered policy concepts were described as follows:

***"Top-down" Design*:** In this option, Congress directs the federal government to identify sites that technical experts determine to be suitable for hosting (nuclear repositories/interim storage facilities) for used nuclear fuel. Federal agencies work with the selected states and local communities to minimize negative economic, environmental, and social impacts while also creating hundreds of jobs and large investments. **This process places priority on technical experts first finding suitable sites, and then working with the affected states and communities to meet their concerns.**

***"Bottom-up" Design*:** In this option, Congress invites states and local communities to apply and compete to host (a nuclear repository/an interim storage facility) for used nuclear fuel that will create hundreds of jobs and large investments. Federal agencies then work with qualified states and communities who want to compete, and the sites that are judged most suitable by technical experts are chosen to host (a nuclear repository/an interim nuclear storage facility). **This process places priority on first finding supportive host communities, then technical experts selecting the most suitable sites among them.**

We then invited participants to rate each conceptual approach. When presented independently, majorities supported each of the concepts, but the bottom-up approach was slightly favored over the top-down approach both by those respondents considering permanent repositories and those considering interim storage facilities.

***B. Institutional Trust and Credibility*:** Another important factor in any design for storing UNF relates to the validity of information provided by government and nongovernmental stakeholders involved in UNF management, and public expectations of associated risk bias. To gain insight into these issues, we asked participants to rate their level of trust in information provided by science and engineering experts from each of nine existing institutions or groups. Because the BRC recommended shifting responsibility for the siting and operation of UNF storage sites from the DOE to a new federally chartered institution, we included one notional entity that we termed "Fedcorp." We then asked respondents to rate their level of trust in risk information from each of the following ten randomly ordered entities:

- The US Nuclear Regulatory Commission
- The US Environmental Protection Agency
- US national laboratories for energy and security
- The National Academy of Sciences (NAS)
- State regulatory agencies
- Environmental advocacy groups, such as the National Resources Defense Council or the Sierra Club
- The Nuclear Energy Institute (NEI), which represents the nuclear power industry
- Utility companies that own nuclear power plants
- The US Department of Energy
- A private company chartered by the government and funded by fees from nuclear energy that is given responsibility for managing used nuclear fuel from US nuclear power plants (Fedcorp) (Appendix C)

We compare results in Figure 5-12.

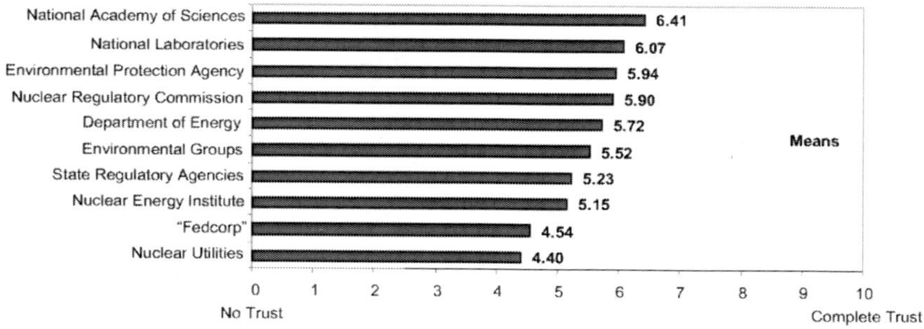

Figure 5-12. Relative Institutional Trust.

Though all but two of the groups were rated above midscale, on average, levels of trust in information from these sources varied significantly, with greatest trust accorded to technical experts from the NAS. Experts from the national laboratories were rated next highest, with those from federal agencies—EPA, NRC, and DOE—receiving modestly high levels of trust, but statistically significantly below those recorded for the National Academy and national laboratories. Mean trust levels expressed in information from advocacy groups, state regulatory agencies, nuclear utilities, and our notional "Fedcorp" entity were significantly lower.

But expectations of the validity of information about nuclear energy issues provided by government and nongovernmental organizations (NGOs) is only part of the process by which members of the public receive and process key elements of information about nuclear energy issues during policy debates. It also is important to understand the filters through which information is processed, and to know what kinds of expected valences members of the public attach to various information sources—especially information about risks associated with nuclear energy and UNF management. To investigate the degree to which systematic bias is expected from technical experts within the above array of organizations and groups involved in nuclear energy and UNF debates, we asked the following:

> Now we want to know more about impressions you may have about how these organizations are likely to assess risks associated with managing radioactive materials, such as used nuclear fuel. Using a scale from one to seven, where one means the organization is likely to *downplay* risks, four means the organization is likely to *accurately assess* risks, and seven means the organization is likely to *exaggerate* risks, please rate your impression of how each organization is likely to assess risks.
>
> [The organizations were presented in random order]

We compare average scores for perceived institutional risk bias in Figure 5-13.

Not surprisingly, the NAS (also the most trusted organization) was perceived to be the least institutionally biased, with a score of 4.10, suggesting a perceived tendency to only slightly overestimate risks. The EPA is perceived to exaggerate risks modestly, and environmental groups to do so more aggressively. All other agencies and groups were believed to downplay risks, with the national laboratories being the least likely to underestimate them. Note that the DOE, the NRC, and state regulatory agencies all were expected to modestly understate risks, while the NEI (which represents the nuclear power industry), our notional entity dubbed "Fedcorp," and nuclear utilities all were judged likely to substantially downplay risks associated with managing UNF.

C. Understanding "Consent": The recommendation by the BRC for a "consent-based" siting process raises important questions, such as what constitutes consent; who should have the authority to grant or withhold consent; and once consent has been legally granted, may it be withdrawn, and, if so, at what stages of the siting process? The BRC considered consent largely to be a function of willingness on the part of authorized entities to enter into legal agreements to host UNF facilities. To gain insight into these questions, we asked participants to rate the importance of the requirement that key stakeholders be required to grant consent during the siting process. With a mean of 7.55 on a scale from zero (not at all important) to ten (extremely important), most respondents judged the consent process to be highly important. We then described in random order two alternative approaches for acquiring consent—one that was more inclusive and one that was less inclusive—as shown below.

More Inclusive: "Consent" should involve a process where many different stakeholders must agree. Thus consent should require agreement by local elected officials, the governor, both US senators, the US congressperson representing the host community, and the state environmental protection agencies. In addition, a state-wide vote should be held that wins the support of a majority of citizens in the host state.

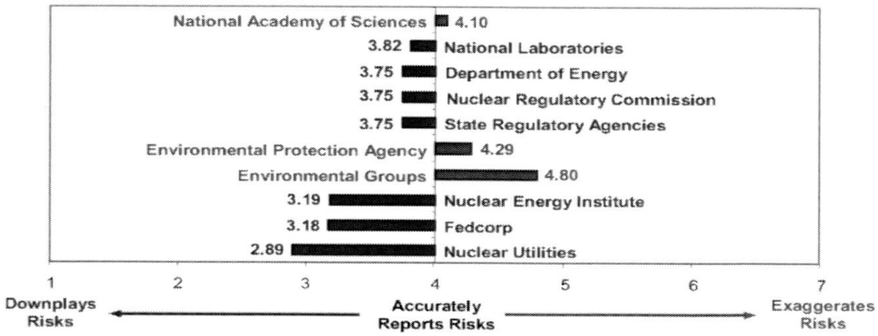

Figure 5-13. Perceived Institutional Risk Bias.

Less Inclusive: "Consent" should involve a process where only those that are most affected must agree. Thus consent should require agreement by local elected officials and the governor. In addition, a vote should be held that wins the support of a majority of the residents in the local host community.

A majority of participants (58%) preferred the more inclusive approach.

Next we asked respondents who they thought should be allowed to block or veto a decision to site a UNF facility (randomly ordered). Responses are summarized in Table 5-4.

Only citizens residing within 50 miles of the proposed site, other state residents, and the host state's governor received majorities of respondent preferences for being able to block or veto a siting decision. While most survey participants preferred a more inclusive process, most also preferred to grant authority to block a siting process only to those citizens most directly affected and to the elected chief executive of the host state.

Our final inquiry into the nature of "consent" for siting UNF facilities dealt with the issue of withdrawing consent after it initially has been granted. To help respondents visualize the long and complex process of siting, we described five stages and then asked participants whether a host state and local community should be allowed to withdraw consent at each of the five sequential stages described below (responses limited to "yes" or "no"). For this process, we employed our same two tracks in which half of the participants were asked to consider withdrawing consent to build a permanent nuclear repository, and the other half of respondents were asked about withdrawing consent for an interim storage site. We show descriptions of the stages and responses in Table 5-5.

Table 5-4. Who Should Be Allowed to Block/Veto a Siting Decision

	% Yes
A majority of the citizens residing within 50 miles of the proposed facilities	68
A majority of the voters of the host state	57
The governor of the host state	55
The host state's environmental protection agency or its equivalent	48
The US Environmental Protection Agency	40
The US Nuclear Regulatory Commission	39
The US Department of Energy	36
The US congressperson representing the district in which the host community is located	35
Either of the two US senators from the host state	34
The leaders of the host state's legislature	29
Nongovernmental environmental groups in the host state	20

Notice that for both permanent repository and interim storage facilities, substantial majorities of respondents thought consent should be allowed to be withdrawn during the first three stages during which site evaluation and licensing is being accomplished. But withdrawing consent in the final two stages of the siting process after a license has been approved did not receive majority support. Essentially, most respondents indicated that once a site license has been issued and construction begins, the time for withdrawing consent has passed.

D. Implications of Proximity: The effects of proximity to UNF storage facilities on attitudes about nuclear facility siting are more complex than might first be assumed. Reactive opposition based on "not- in-my-backyard" (NIMBY) or "locally unwanted land use" (LULU) concerns is a common assumption (Groothuis and Miller 1994; Kraft and Clary 1991). But the actual relationship between proximity and public support or opposition to nuclear facilities has been more subtle and complex (Greenberg 2009; Jenkins-Smith et al. 2011). For example, near proximity can be associated with increased jobs, improved roads, and economic benefits that dampen traditional NIMBY assumptions. Familiarity with the proposed facility, and with those who work there, may also dampen NIMBY responses. BRC recommendations for consent-based siting processes involving volunteer host communities raise at least three relevant questions about proximity. What, if any, are the implications of residents' proximity to *current* temporary storage sites for their support of or opposition to a national strategy of moving UNF from those temporary sites to centralized permanent repositories and longer-term interim storage facilities? How does proximity to proposed (and hypothetical) new consolidated UNF facilities influence support or opposition, and how do those effects vary with distances? And does the experience of having lived without adverse effects near an existing UNF temporary storage site condition acceptance of future consolidated facilities near one's residence (Jenkins-Smith et al. 2012b)?

To address these dimensions of proximity, we estimated the straight-line distance from the primary residence of each respondent to the nearest facility at which UNF currently is stored.[28] We estimate that about three out of four US residents reside with 100 miles of a licensed UNF storage facility, and our respondents were similarly situated, averaging 73.3 miles. Figure 5-14 shows the locations of our respondents (each represented by a dot) in relation to current UNF storage sites (shown as triangles).

Table 5-5. Withdrawing Consent for UNF Facilities

	Permanent Repository % Yes	Interim Storage % Yes
Stage 1: The community or state volunteers to be a candidate to host (a permanent repository/an interim storage facility) for used nuclear fuel, and a technical evaluation of the site is begun. This evaluation may take several years to complete. Should the host state and local community be allowed to withdraw their consent during this stage?	74.3	74.7
Stage 2: Scientific evaluation of the suitability of the site for (permanent storage and disposal/interim storage) of used nuclear fuel is completed. Should the host state and local community be allowed to withdraw their consent at this stage?	73.6	72.6
Stage 3: If the site is determined to be suitable, a license to construct (a permanent repository/an interim storage facility) for used nuclear fuel is submitted to the US regulatory agencies; the regulatory consideration may take several years to complete. Should the host state and local community be allowed to withdraw their consent during this stage?	62.2	66.2
Stage 4: If the license is provided, construction of (a permanent repository/an interim storage facility) for used nuclear fuel begins. Construction will take several years to complete. Should the host state and local community be allowed to withdraw their consent during this stage?	40.5	46.3
State 5: Construction is completed, and the (permanent repository/interim storage facility) is prepared to receive used nuclear fuel. Should the host state and local community be allowed to withdraw their consent at this stage?	30.5	33.9

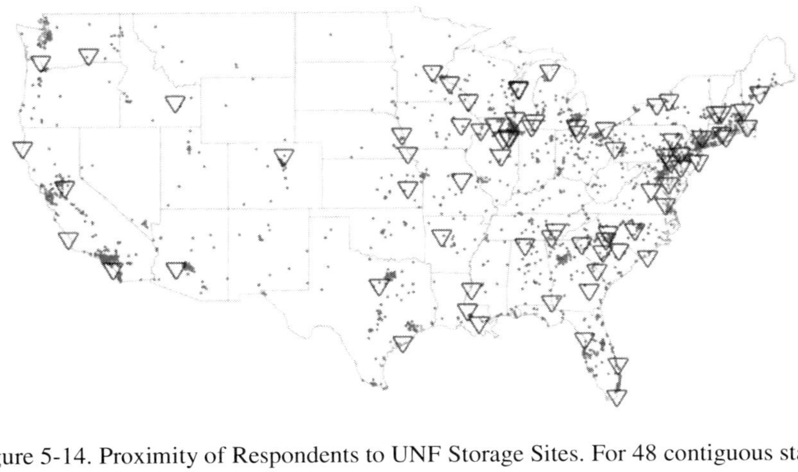

Figure 5-14. Proximity of Respondents to UNF Storage Sites. For 48 contiguous states, 75.6% of population resides within 100 miles of current on-site storage of UNF and 44.0% reside within 50 miles.

To determine if current proximity to UNF facilities influenced support for building permanent repositories or interim storage facilities, we again used the dual tracks previously described. Employing ordinary least-squares regressions with estimated proximity to current sites as the independent variable, and support for either permanent repositories or interim storage facilities as the dependent variable, we determined that support for either type of new storage sites was not systematically related to the respondent's current proximity to temporary UNF facilities. These findings suggest that one's proximity to a current UNF facility is not predictive of a general preference for a national strategy for the future management of UNF—whether the strategy is permanent repositories or interim storage facilities or a combination.

To more directly address NIMBY assumptions about new UNF facilities, we asked respondents to rate their support or opposition for the same two types of facilities if their primary residence was located within one of three randomly assigned distances: 300 miles, 100 miles, or 50 miles. Table 5-6 compares mean support for each of the two types of facilities given no assumptions of proximity and, comparatively, with assumptions of progressive proximity to the respondent's home.

As Table 5-6 illustrates, mean support for either type of new UNF facility declined significantly with increasing proximity. At 50 miles assumed proximity, support for permanent repositories declined about 23%, while support for interim facilities at the same assumed distance decreased about 14%. These data suggest that NIMBY is at work, and that when considering how to site consolidated facilities, support initially can be expected to decrease with distance from the proposed facilities. However, experience in siting the WIPP in southern New Mexico revealed nonlinear variation in support versus proximity, with support increasing appreciably within the localized zone of greatest perceived economic and infrastructure benefits and enhanced emergency response capabilities (Jenkins-Smith et al. 2011).

To address our third proximity question about whether living near a current UNF facility conditions support or opposition to proposed new consolidated UNF storage facilities, we used the information in respondents' proximity to existing UNF storage to conduct an experiment: one group of randomly selected respondents were not informed of their proximity to current UNF storage; the remainder where provided with an estimate of the linear distance between their residences and the nearest UNF storage facility, and were offered a map showing the location of UNF storage facilities nationwide. For purposes of this analysis, we sorted the survey respondents into one of three groups: (1) those who were not shown their proximity to UNF storage;

(2) those who were shown that their primary residence is within 25 miles of a UNF storage facility, and (3) those who were shown that their primary residence was located farther than 25 miles from a UNF storage facility. Tables 5-7 and 5-8 show average levels of support for consolidated UNF storage facilities within 50, 100, and 300 miles of each respondent's primary residence.

Our respondents were moderately more willing to accept new UNF facilities (of either type) if they were aware that they currently reside within 25 miles of an existing UNF site. In Table 5-7, those who had been informed that they reside within 25 miles of UNF reported nominally higher support than those who either resided at greater distances or had not been informed of their proximity to an existing site. This increase in support was statistically significant when the new repository was to be approximately 100 miles away. The level of support of those not informed of proximity was similar to that of respondents residing at distances greater than 25 miles from existing sites. Results for siting an interim storage facility (Table 5-8) were similar, except that support for an interim facility increased among those informed of proximity (regardless of distance) from existing UNF sites.

Table 5-6. Mean Support for Future UNF Sites by Assumed Proximity

	No Proximity Specified	300 Miles from Residence	100 Miles from Residence	50 Miles from Residence
Permanent repositories	4.65	4.35	4.00	3.60
Interim storage facilities	4.29	4.25	3.87	3.74

Table 5-7. Effects of Current Proximity on Support for a Future Permanent Repository

Distance from Home to Current UNF Site	50 Miles from Home	Support When Site is 100 Miles from Home	300 Miles from Home
Shown to be within 25 miles	3.89	4.79	4.71
Shown to be over 25 miles	3.56	3.90	4.33
Distance not shown	3.55	3.96	4.26
Model *F* statistic significance	not significant	$p = 0.01$	not significant
Sample size	330	311	367

Table 5-8. Effects of Current Proximity on Support for an Interim Storage Facility

Distance from Home to Current UNF Site	50 Miles from Home	Support When Site is 100 Miles from Home	300 Miles from Home
Shown to be within 25 miles	4.34	4.12	4.22
Shown to be over 25 miles	3.71	3.97	4.30
Distance not shown	3.42	3.49	4.13
Model *F* statistic significance	$p = 0.02$	$p = 0.09$	not significant
Sample size	289	344	366

5.3. Summary

These survey findings should be taken as tentative because they are of the general population rather than representative of a potential host community, but they provide evidence that proximity to current and future UNF facilities and the siting of either a permanent repository or an interim storage site is a complex relationship that does not necessarily conform to NIMBY expectations. We find that the effects of proximity can be subtle, subject to conditioning, and nonlinear. However, these data do suggest that potential host communities whose residents have no experience living near UNF may be less receptive than communities whose citizens have lived near such facilities. Our preliminary finding is that knowledge of the proximity of *existing* UNF sites to a potential host community may have systematic effects on support for *new* UNF facilities of either type.

Summary and Implications

This report has provided findings from a set of on-going social science studies undertaken by the University of Oklahoma's Center for Risk and Crisis Management (CRCM) in collaboration with Sandia National Laboratories (SNL). These studies test some of the wide-spread assumptions about the conditions under which nuclear facility siting does and does not work; analyze when and how major nuclear events (like that in Fukushima, Japan) focus public attention and reshape public understanding and support for nuclear facilities; and evaluate how the design features of siting programs can facilitate the legitimacy of and support for a siting program among the United States (US) public.

The combination of studies described here provide a broad, empirically grounded assessment of past nuclear facility siting efforts, the changing long-term patterns of public perceptions of the risks posed by nuclear energy in the US, and an in-depth analysis of current American preferences for the design of fair and effective processes for siting used nuclear fuel (UNF) storage and disposal facilities. They also provide guidance for the kinds of future research that will most assist in the design and development consent-based nuclear facility siting programs of the kind recommended by the Blue Ribbon Commission on America's Nuclear Future (BRC 2012) and endorsed in the DOE response (DOE 2013).

The first study, described in Section 2 of the report, is a comparative analysis of nuclear facility siting cases, coded for key characteristics of the siting country, context and program. It provides an opportunity to analyze the set of circumstances in which a proposed nuclear facility is more, or less, likely to be sited and made operational. Using 269 cases, based in recognized efforts to site nuclear facilities in countries across the globe, this study employed quantitative analysis to test propositions about the factors that change the probability of siting outcomes.

The results of this comparative case analysis shows that the probability that a proposed nuclear facility will become operational has (a) decreased substantially over time, and (b) is more difficult (less likely) when siting in more democratic, federal-type governmental systems, even when including other variables such as the recent occurrence of nuclear crises. Why do countries with greater democratic openness, as defined by the widely used Polity scores, have a more difficult time siting nuclear facilities? A more detailed analysis shows that greater democratic openness is associated an increased probability of expressed opposition to the facility that, in turn, diminishes the probability that the facility will be sited. In addition, the inclusion of mechanisms for public involvement in siting programs is highly associated with expressed opposition, but, as they were employed in the past, had no bearing on the likely outcome of the siting effort.

The second study, as reported in Section 3, uses evidence from a very large array of surveys taken over time, in combination with an innovative data aggregation technique, to analyze fluctuations and trends in public opinion on nuclear issues over a long time horizon. Two data series – representing perceived nuclear risks and the acceptability of nuclear power – demonstrated the potent and consistent effect of nuclear crisis events (Three Mile Island—

TMI, Chernobyl, and Fukushima) on public opinion; the effects of these events are durable and cumulative. These kinds of events have resulted in sizable and sustained declines in support for nuclear energy. This analysis also identifies two important underlying trends in public opinion concerning nuclear issues; perceived nuclear risks have trended downward (at a declining rate) since the mid-1970s, but over the same period broader public support for nuclear energy has also declined. These results are consistent with our finding in Section 2 that the percentage of proposed nuclear facilities that have been cancelled before becoming operational has grown over time.

The third study (Section 4) shows new ways to track and analyze shifts in public attention on issues like nuclear energy and nuclear waste management, using new types of data streams that provide remarkable, real-time pictures of the shifts in public attention to nuclear issues before, during and after the Fukushima event. The analysis of two streams – Twitter traffic and Google searches – shows a large pulse in attention to nuclear issues following the event, followed by a sustained increase in attention to nuclear issues. The largest proportionate increase in attention was evident in regions that had experience, of one kind or another, with nuclear energy and/or nuclear waste facility siting. Overall the Fukushima event seems to have established a "new normal" level of attention to nuclear issues, with the highest levels apparent in regions with past nuclear experience.

The fourth study (Section 5) evaluates the current and future public willingness to support consent based siting efforts for nuclear facilities, based on a large (n = 2017) nationwide Internet survey of adult respondents that demographically match the US Census. The survey first provided context by evaluating the relative concern about energy issues, preferences across different energy sources, and support for nuclear energy. While respondents were optimistic about the availability energy supplies in the future, they were deeply concerned about over-reliance on insecure foreign sources of supply. Environmental concerns about climate change, while related to support for fossil fuels, did not increase support for nuclear energy. Overall, support for new nuclear reactors was mixed.

Our analysis of support for nuclear energy found, not surprisingly, that beliefs about risks and benefits of nuclear energy were the most powerful predictors of support for new reactors. Support for nuclear energy was also increased by trust in the federal government and – in particular – trust in responsible federal agencies to provide information about nuclear energy. Consistent with the findings of the studies reported in Sections 3 and 4, the Fukushima event reduced support for reliance on nuclear power in the US.

After establishing context, the survey focus shifted to perceptions and preferences concerning the management of UNF. Respondents preferred permanent repositories and/or interim storage facilities over current temporary on-site storage practices, but the rationale and benefits of permanent repositories seemed more persuasive than those for interim storage facilities. Support for either permanent repositories or interim storage facilities varied with physical design characteristics and financial compensation (beyond associated jobs and indirect economic benefits), and these data strongly suggest that tailoring facility concepts, designs, and incentive structures to local conditions may potentially influence public support for or opposition to siting options.

When respondents are asked about how a facility siting program should be structured, a "bottom-up approach" emphasizing volunteer communities and states offering to be considered for siting in return for benefits was slightly favored over the top-down approach in which communities with technically optimal sites are identified and approached by siting agencies. Public trust for the agencies and groups likely to be prominently involved in the siting process varied substantially; not surprisingly, the National Academy of Sciences (NAS) was most trusted and was perceived to be the least institutionally biased. Federal agencies (the Nuclear Regulatory Commission (NRC), Department of Energy (DOE) and the Environmental Protection Agency (EPA)) were generally perceived to be trustworthy, and to have only modest tendencies to overstate (EPA) or understate (NRC and DOE) the risks posed by nuclear facilities. Environmental nongovernmental organizations (NGOs) were expected to aggressively overstate risks to the public, while industry groups and a hypothetical governmentally-chartered private entity (Fedcorp) were all seen as likely to substantially understate the risks.

Consent-based siting implies an ability to withdraw that consent, but at what point in the process can consent no longer be withdrawn? Sizable majorities of the survey respondents thought candidate communities and states should be able to withdraw consent during the early stages of the siting process (site evaluation and licensing), but not in the final two stages (construction and operation) of the process after licensing. Essentially, most respondents indicated that once a site license has been issued and construction begins, the time for withdrawing consent has passed.

The most challenging issue for consent-based siting will be obtaining support from residents in or near the host community, so the survey carefully analyzed how proximity to a proposed facility would affect support for siting. Of particular interest was the link between proximity to existing UNF storage and support for new facilities. We found that the effects of proximity can be subtle. Our findings suggest that knowledge of the proximity of *existing* UNF sites to a potential host community may increase support for *new* UNF facilities (76% of the US population lives within 100 miles and 44% live within 50 miles of current on-site storage of UNF). We note that these data, while based on a large nationwide sample and a hypothetical siting case, are indicative of the responses that will be obtained when a specific proposal is considered by an identified potential community. Furthermore, the influence of knowledge about proximity to existing facilities is consistent with regional surveys around existing facilities as opposed to questions about hypothetical facilities.

As a group, these studies provide strong support for the need to redefine our approach to UNF facility siting. The traditional, top-down design in which technical considerations are used to define a site and relevant communities subsequently approached for consent (or acquiescence) has been of declining efficacy for decades. Periodic nuclear crisis events have added a series of negative punctuations, adding a further burden to siting efforts. These events shift public attention in systematic ways, and alter long-term scrutiny by the public to nuclear events. It is worth noting that, as historically practiced, the nuclear facility siting initiatives have been most effective when the public has had *less* opportunity to participate. Historical modes of public involvement – typically in the form of hearings and public information programs – appears to have been in response to opposition, and appear to have no net effect on the ultimate outcome of the siting process. In the modern world, as communities and nations increasingly opt for greater public access, involvement, and self-determination, the old pattern of public involvement and siting is increasingly obsolete. The data speak clearly to this trend. In short, new approaches and strategies are needed in the context of increasing participatory opportunities.

These findings are consistent with the conclusions of the BRC that in the US, in order to site UNF storage and disposal facilities, it will be necessary to alter the traditional mode of the facility siting process from a top-down approach to one that relies on a consent-based program, that provides host communities and states a prominent role, and that recognizes that the features of both the siting process and the facility design can have substantial bearing on public support. The survey data provide ample evidence that a UNF storage facility – be it a permanent repository or an interim site – are more acceptable when they are not only safe but provide benefits that offset the risks (research and development), that consider the possible future resource value of the UNF, and that promise to provide long-term community resources. While our analysis indicates that "not in my back yard" (NIMBY)-like responses will be forthcoming, these responses are mitigated when respondents realize that they reside within a reasonably short distance of an existing UNF storage sites. The data confirm that appropriately designed siting programs, reflecting widely-held norms respecting the right of communities to consent (or withdraw), and that permit flexibility on the design and function of the management facilities will have substantially greater support than traditional, single-function facilities employing top-down siting strategies.

REFERENCES

Atkinson, Mary Layton., et al. (2011a). "Developing Policy-Specific Conceptions of Mood: The United States." In Annual Meeting of the Comparative Agendas Project, http://www.unc.edu/~fbaum/papers/Equality-Mood-Catania- June-2011.pdf.

Atkinson, Mary Layton., et al. (2011b). "Mood and Agendas: Developing Policy-Specific Conceptions of Mood." In Annual Meeting of the Midwest Political Science Association, http://www.unc.edu/~fbaum/papers/MPSA-2011- Mood-Policy.pdf.

BRC (Blue Ribbon Commission on America's Nuclear Future). (2012). Report to the Secretary of Energy. Available online at: http://cybercemetery.unt.edu/archive/brc/20120620220235/http://brc.gov/sites/default/files/documents/brc_finalreport_jan2012.pdf.

Brulle, Robert, J., Jason, Carmichael. & Craig Jenkins, J. (2012). "Shifting Public Opinion on Climate Change: An Empirical Assessment of Factors Influencing Concern over Climate Change in the U.S., 2002–2010." Climatic Change. http://www.springerlink.com/content/ k17856khp026w174/(Accessed February 10, 2012).

DOE (US Department of Energy). (2013). *Strategy for the Management and Disposal of Used Nuclear Fuel and High- Level Radioactive Waste.* Available at www.doe.gov.

Douglas, Mary. (1970). *Natural Symbols: Explorations in Cosmology.* London: Barrie and Rockliff.

Douglas, Mary. & Aaron, Wildavsky. (1982). *Risk and Culture: An Essay on the Selection of Technical and Environmental Dangers.* Berkeley, CA: University of California Press.

Downs, Anthony. (1957). *An Economic Theory of Democracy.* New York: Harper.

Eckstein, Harry. (1975). *Patterns of Authority: A Structural Basis for Political Inquiry*. New York: John Wiley & Sons.

Erikson, Robert S., Michael B. Mackuen & James, A. Stimson. (2002). The Macro Polity. Cambridge University Press.

Fiske, Susan T. & Shelly, E. Taylor. (1992). *Social Cognition.* New York: McGraw-Hill.

Herron, Kerry G. & Hank, C. Jenkins-Smith. (2002). "U.S. Perceptions of Nuclear Security in the Wake of the Cold War: Comparing Public and Elite Belief Systems." *International Studies Quarterly*, *46*(4), 450–479.

Herron, Kerry G. & Hank, C. Jenkins-Smith. (2006). *Critical Masses and Critical Choices: Evolving Public Opinion on Nuclear Weapons, Terrorism, and Security.* Pittsburgh, PA: University of Pittsburgh Press.

Hurwitz, Jon. & Mark, Peffley. (1987). "How are Foreign Policy Attitudes Structured? A Hierarchical Model." *American Political Science Review*, *8*(4), 1099–1120.

Hurwitz, Jon, Mark Peffley & Mitchell, A. Seligson. (1993). "Foreign Policy Belief Systems in Comparative Perspective: The United States and Costa Rica." *International Studies Quarterly*, *37*(3), 245–270.

Jenkins-Smith, Hank C., Kerry, Herron. & Carol, Silva. (2011). *Perspectives on Energy, the Environment, Nuclear Weapons, and Terrorism: 2010 National Security Surveys.* Sandia Report: SAND2011-1686, Albuquerque, NM: Sandia National Laboratories.

Jenkins-Smith, Hank C. & Kerry, G. Herron. (2002a). *Comparing Public Views on Security: U.S. National Security Surveys, 1993–2002. Volume I: U.S./British Public Views on Nuclear Weapons and Nuclear Energy.* Sandia Report: SAND2002-1187P. Albuquerque, NM: Sandia National Laboratories.

Jenkins-Smith, Hank C. & Kerry, G. Herron. (2002b). *Comparing Public Views on Security: U.S. National Security Surveys, 1993–2002. Volume II: Trends in U.S. Perspectives on Nuclear Security, Terrorism, and Energy.* Sandia Report: SAND2002-2401P. Albuquerque, NM: Sandia National Laboratories.

Jenkins-Smith, Hank C. & Kerry, G. Herron. (2005). "United States Public Response to Terrorism: Fault Lines or Bedrock?" *Review of Policy Research*, *22*(5), 599–623.

Jenkins-Smith, Hank C. & Howard, Kunreuther. (2011). "Mitigation and Benefits Measures as Policy Tools for Siting Potentially Hazardous Facilities: Determinants of Effectiveness and Appropriateness," *Risk Analysis*, *21*, 371-382.

Jenkins-Smith, Hank C., Neil, J. Mitchell. & Kerry, G. Herron. (2004). "Foreign and Domestic Policy Belief Structures in the U.S. and Britain." *Journal of Conflict Resolution*, *48*(3), 287–309.

Jenkins-Smith, Hank C., Carol, L. Silva., Kerry, G. Herron., Sarah, R. Trousset. & Rob, P. Rechard. (2012a). "Enhancing the Acceptability and Credibility of a Repository for Spent Nuclear Fuel." *The Bridge*, *42*(2), 49–58.

Jenkins-Smith, Hank C., Carol, L. Silva., Kerry, G. Herron., Evaristo, "Tito" Bonano. & Rob, P. Rechard. (2012b). "Designing a Process for Consent-Based Siting of Used Nuclear Fuel Facilities: Analysis of Public Support." *The Bridge*, *42*(3), 28–39.

Jenkins-Smith, Hank C. & Walter, K. Smith. (1994). "Ideology, Culture, and Risk Perception." In *Politics, Policy and Culture*, eds. Dennis J. Coyle and Richard J. Ellis. Boulder, CO: Westview Press.

Kasperson, Roger E. et al. (1988). "The Social Amplification of Risk: A Conceptual Framework." *Risk Analysis*, *8*(2), 177–187.

Kasperson, Roger E., Nayna, Jhaveri & Jeanne, X. Kasperson. (2001). "Stigma and the Social Amplification of Risk." In Risk, Media and Stigma: Understanding Public Challenges to Modern Science and Technology. London: Earthscan Publications Ltd, eds. James Flynn, Paul Slovic, and Howard Kunreuther. London, UK: Earthscan Publications, p. 9–27.

Mahaffery, James. (2009). *Atomic Awakening: A New Look at the History and Future of Nuclear Power*. New York: Pegasus.

Page, Benjamin. & Robert, Shapiro. (1983). "Effects of Public Opinion on Policy." *American Political Science Review*, *77*(1), 175-190.

Peffley, Mark A. & Jon, Hurwitz. (1985). "A Hierarchical Model of Attitude Constraint." *American Journal of Political Science*, *29*(4), 871–890.

Sabatier, Paul A. & Hank, C. Jenkins-Smith. (1993). *Policy Change and Learning: An Advocacy Coalition Approach.* Boulder, CO: Westview Press.

Sabatier, Paul A. & Hank, C. Jenkins-Smith. (1999). "The Advocacy Coalition Framework: An Assessment." In *Theories of the Policy Process*, eds. Paul A. Sabatier and Hank C. Jenkins-Smith. Boulder, CO: Westview Press.

Schattschneider, Elmer Eric. (1960). *The Semisovereign People: A Realist's View of Democracy in America*. Holt, Rinehart and Winston.

Slovic, Paul., Mark, Layman. & James, H. Flynn. (1980). *What Comes To Mind When You Hear The Words "Nuclear Waste Repository"? A Study Of 10,000 Images*. Carson City, NV: Nevada Agency for Nuclear Projects, Yucca Mountain Socioeconomic Project.

Stimson, James A. (1999). Public Opinion in America: Moods, Cycles, and Swings. 2nd ed. Boulder, CO: Westview Press.
Stimson, James A., Michael, B. Mackuen & Robert, S. Erikson. (1995). "Dynamic Representation." *The American Political Science Review*, *89*(3), 543–565.
Thompson, Michael. & Aaron, Wildavsky. (1982). "A Proposal to Create a Cultural Theory of Risk." In *The Risk Analysis Controversy: An Institutional Perspective.* Eds. Howard C. Kunreuther and E. V. Ley. New York: Springer- Verlag.
Weart, Spencer. (2012). *The Rise of Nuclear Fear*. Harvard University Press.
Wildavsky, Aaron and Karl Dake. 1990. "Theories of Risk Perception: Who Fears What and Why?" *Daedalus*, *119*(4), 41–60.

Appendix A: Cases and References for Nuclear Facility Siting

A.1. List of Cases

Name of Facility	Country	Year of Decision	Outcome
Atucha I Nuclear Power Plant	Argentina	1974	Completed
Embalse Nuclear Power Plant	Argentina	1983	Completed
Metsamor Nuclear Power Plant*	Armenia (USSR)	1976	Completed
Zwentendorf Nuclear Power Plant	Austria	1978	Cancelled
Doel Nuclear Power Station	Belgium	1974	Completed
Tihange Nuclear Power Station	Belgium	1975	Completed
Angra Nuclear Power Plant*	Brazil	1982	Completed
Belene Nuclear Power Plant	Bulgaria	2012	Cancelled
Kozloduy Nuclear Power Plant*	Bulgaria	1974	Completed
Bruce Nuclear Generating Station	Canada	1976	Completed
Darlington Nuclear Generating Station	Canada	1990	Completed
Gentilly-I Nuclear Generating Station	Canada	1970	Completed
Pickering Nuclear Generating Station	Canada	1971	Completed

Name of Facility	Country	Year of Decision	Outcome
Point Lepreau Nuclear Generating Station	Canada	1982	Completed
Temelin Nuclear Power Station	Czech Republic	2000	Completed
Dukovany Nuclear Power Station	Czech Republic (Czechoslovakia)	1985	Completed
Loviisa Nuclear Power Plant	Finland	1977	Completed
Olkiluoto Nuclear Power Plant	Finland	1978	Completed
Belleville Nuclear Power Plant	France	1987	Completed
Blayais Nuclear Power Plant*	France	1981	Completed
Brennilis Nuclear Power Plant	France	1966	Completed
Bugey Nuclear Power Plant	France	1972	Completed
Cattenom Nuclear Power Plant	France	1986	Completed
Chinon Nuclear Power Plant	France	1962	Completed
Chooz Nuclear Power Plant	France	1966	Completed
Civaux Nuclear Power Plant	France	1997	Completed
Cruas Nuclear Power Plant	France	1983	Completed
Dampierre Nuclear Power Plant	France	1980	Completed
Fessenheim Nuclear Power Plant	France	1977	Completed
Flamanville Nuclear Power Plant*	France	1985	Completed
Golfech Nuclear Power Plant*	France	1990	Completed
Gravenlines Nuclear Power Plant	France	1980	Completed
Le Carnet Nuclear Power Plant	France	1997	Cancelled
Marcoule Nuclear Site	France	1958	Completed
Nogent Nuclear Power Plant	France	1987	Completed
Paluel Nuclear Power Plant	France	1984	Completed
Penly Nuclear Power Plant	France	1990	Completed
Plogoff Nuclear Power Plant	France	1981	Cancelled
Saint-Alban Nuclear Power Plant*	France	1985	Completed
Saint-Laurent Nuclear Power Plant*	France	1969	Completed
Superphenix Nuclear Power Plant	France	1985	Completed
Tricastin Nuclear Power Plant	France	1980	Completed
Stendal Nuclear Power Plant	Germany	1990	Cancelled
Greifswald nuclear power station*	Germany (East)	1973	Completed

(Continued)

Name of Facility	Country	Year of Decision	Outcome
Rheinsberg Nuclear Power Station	Germany (East)	1966	Completed
Wurgassen Nuclear Power Plant	Germany (East)	1971	Completed
Biblis Nuclear Power Plant	Germany (West)	1974	Completed
Breisach Nuclear Power Plant	Germany (West)	1973	Cancelled
Brokdorf Nuclear Power Plant	Germany (West)	1986	Completed
Brunsbuttel Nuclear Power Plant	Germany (West)	1976	Completed
Emsland Nuclear Power Plant	Germany (West)	1988	Completed
Grafenrheinfeld Nuclear Power Plant	Germany (West)	1981	Completed
Grohnde Nuclear Power Plant	Germany (West)	1984	Completed
Gundremmingen Nuclear Power Plant	Germany (West)	1966	Completed
Isar Nuclear Power Plant	Germany (West)	1977	Completed
Kruemmel Nuclear Power Plant*	Germany (West)	1983	Completed
Mulheim-Karlich Nuclear Power Plant	Germany (West)	1986	Completed
Neckarwestheim Nuclear Power Station	Germany (West)	1976	Completed
Obrigheim Nuclear Power Plant	Germany (West)	1968	Completed
Philippsburg Nuclear Power Plant	Germany (West)	1979	Completed
SNR-300 Fast Breeder Reactor*	Germany (West)	1995	Cancelled
Stade Nuclear Power Plant	Germany (West)	1972	Completed
Unterweser Nuclear Power Plant	Germany (West)	1978	Completed

Name of Facility	Country	Year of Decision	Outcome
Wyhl Nuclear Power Plant	Germany (West)	1977	Cancelled
Paks Nuclear Power Plant	Hungary	1982	Completed
Haripur Nuclear Power Plant	India	2011	Cancelled
Kaiga Atomic Power Station	India	2000	Completed
Kakrapar Atomic Power Station	India	1992	Completed
Madras Atomic Power Station	India	1983	Completed
Narora Atomic Power Station	India	1989	Completed
Rajasthan Atomic Power Station*	India	1972	Completed
Tarapur Atomic Power Station	India	1969	Completed
Caorso Nuclear Power Plant	Italy	1977	Completed
Enrico Fermi (Trino Vercellese) Nuclear Power Plant*	Italy	1964	Completed
Garigliano Nuclear Power Plant	Italy	1963	Completed
Latina Nuclear Power Plant	Italy	1962	Completed
Montalto di Castro (Alto Lazio) Nuclear Power Plant*	Italy	1987	Cancelled
Fukushima Daiichi Nuclear Power Plant	Japan	1970	Completed
Genkai Nuclear Power Plant	Japan	1975	Completed
Hamaoka Nuclear Power Plant	Japan	1974	Completed
Higashidori Nuclear Power Plant	Japan	2005	Completed
Ikata Nuclear Power Plant	Japan	1977	Completed
Kashiwazaki-Kariwa Nuclear Power Plant*	Japan	1984	Completed
Maki Nuclear Power Plant*	Japan	2004	Cancelled
Mihama Nuclear Power Plant	Japan	1970	Completed
Monju Nuclear Power Plant	Japan	1994	Completed
Ohi Nuclear Power Plant	Japan	1977	Completed
Onagawa Nuclear Power Plant	Japan	1983	Completed
Sendai Nuclear Power Plant	Japan	1983	Completed
Shika Nuclear Power Plant	Japan	1992	Completed
Shimane Nuclear Power Plant	Japan	1973	Completed
Suzu Nuclear Power Plant	Japan	2003	Cancelled
Takahama Nuclear Power Plant	Japan	1974	Completed
Tokai Nuclear Power Plant	Japan	1965	Completed
Tomari Nuclear Power Plant*	Japan	1988	Completed

(Continued)

Name of Facility	Country	Year of Decision	Outcome
Tsuruga Nuclear Power Plant	Japan	1969	Completed
Kori Nuclear Power Plant	Korea	1977	Completed
Uljin Nuclear Power Plant*	Korea	1988	Completed
Wolseong Nuclear Power Plant*	Korea	1982	Completed
Yeonggwang/Youngkwang Nuclear Power Plant	Korea	1986	Completed
Ignalina Nuclear Power Plant	Lithuania (USSR)	1983	Completed
Laguna Verde Nuclear Power Plant	Mexico	1988	Completed
Borssele Nuclear Power Plant*	Netherlands	1973	Completed
Dodewaard Nuclear Power Plant	Netherlands	1968	Completed
Chashma Nuclear Power Plant	Pakistan	2000	Completed
Karachi Nuclear Power Plant	Pakistan	1971	Completed
Cernavoda Nuclear Power Plant	Romania	1996	Completed
Bohunice Nuclear Power Plant*	Slovakia (Czechoslovakia)	1972	Completed
Mochovce Nuclear Power Plant	Slovakia (Slovac Republic)	1998	Completed
Krsko Nuclear Power Plant	Slovenia (Yugoslovia)	1981	Completed
Koeberg Nuclear Power Station	South Africa	1984	Completed
Almaraz Nuclear Power Plant	Spain	1981	Completed
Asco Nuclear Power Plant	Spain	1983	Completed
Cofrentes Nuclear Power Plant*	Spain	1984	Completed
Jose Cabrera Nuclear Power Station	Spain	1968	Completed
Lemoniz Nuclear Power Plant	Spain	1983	Cancelled
Santa Maria de Garona Nuclear Power Plant	Spain	1970	Completed
Sayago Nuclear Plant*	Spain	1983	Cancelled
Trillo Nuclear Power Plant	Spain	1988	Completed
Valde Caballeros Nuclear Power Plant	Spain	1983	Cancelled
Vandellos Nuclear Power Plant	Spain	1972	Completed
Barseback Nuclear Power Plant	Sweden	1975	Completed
Brodalen Nuclear Power Plant	Sweden	1980	Cancelled

Name of Facility	Country	Year of Decision	Outcome
Forsmark Nuclear Power Plant	Sweden	1980	Completed
Oskarshamn Nuclear Power Plant*	Sweden	1970	Completed
Ringhals Nuclear Power Plant	Sweden	1973	Completed
Beznau Nuclear Power Plant	Switzerland	1969	Completed
Gosgen Nuclear Power Plant*	Switzerland	1979	Completed
Kaiseraugst Nuclear Power Plant	Switzerland	1990	Cancelled
Leibstadt Nuclear Power Plant	Switzerland	1984	Completed
Muhleberg Nuclear Power Plant	Switzerland	1971	Completed
Niederamt Nuclear Power Plant	Switzerland	2011	Cancelled
Akkuyu Nuclear Power Plant	Turkey	2012	Cancelled
Chernobyl Nuclear Power Plant	Ukraine (USSR)	1977	Completed
Khmelnytsky Nuclear Power Plant	Ukraine (USSR)	1987	Completed
Rivne Nuclear Power Plant	Ukraine (USSR)	1980	Completed
South Ukraine Nuclear Power Plant	Ukraine (USSR)	1982	Completed
Zaporizhia Nuclear Power Plant	Ukraine (USSR)	1984	Completed
Berkeley Nuclear Power Station	United Kingdom	1961	Completed
Bradwell Nuclear Power Station	United Kingdom	1961	Completed
Braystones Nuclear Power Station	United Kingdom	2010	Cancelled
Calder Hall Nuclear Power Station	United Kingdom	1956	Completed
Chapelcross Nuclear Power Station	United Kingdom	1958	Completed
Dungeness Nuclear Power Station	United Kingdom	1965	Completed
Hartlepool Nuclear Power Station*	United Kingdom	1983	Completed
Heysham Nuclear Power Station	United Kingdom	1983	Completed
Hinkley Point Nuclear Power Station*	United Kingdom	1964	Completed
Hunterston Nuclear Power Station*	United Kingdom	1963	Completed
Kirksanton Nuclear Power Station	United Kingdom	2010	Cancelled
Oldbury Nuclear Power Station	United Kingdom	1967	Completed

(Continued)

Name of Facility	Country	Year of Decision	Outcome
Sizewell Nuclear Power Station	United Kingdom	1965	Completed
Torness Nuclear Power Station*	United Kingdom	1988	Completed
Trawsfynydd Nuclear Power Station	United Kingdom	1964	Completed
Wylfa Nuclear Power Station	United Kingdom	1969	Completed
Alan R. Barton Plant	USA	1975	Cancelled
Allens Creek Nuclear Power Plant	USA	1982	Cancelled
Arkansas Nuclear One	USA	1974	Completed
Atlantic Nuclear Power Plant*	USA	1978	Cancelled
Bailly Nuclear Power Plant	USA	1981	Cancelled
Beaver Valley Power Station*	USA	1976	Completed
Bellefonte Nuclear Generating Station	USA	1988	Cancelled
Big Rock Point Power Plant	USA	1962	Completed
Black Fox Nuclear Power Plant	USA	1982	Cancelled
Blue Hills Nuclear Power Plant	USA	1978	Cancelled
Bodega Bay Nuclear Power Plant	USA	1964	Cancelled
Braidwood Station	USA	1987	Completed
Browns Ferry Nuclear Plant	USA	1974	Completed
Brunswick Steam Electric Plant	USA	1975	Completed
Byron Station	USA	1985	Completed
Callaway Plant	USA	1984	Completed
Calvert Cliffs Nuclear Power Plant	USA	1975	Completed
Carroll County Nuclear Power Plant*	USA	1988	Cancelled
Catawba Nuclear Station	USA	1985	Completed
Cherokee Nuclear Power Plant	USA	1983	Cancelled
Clinton Power Station	USA	1987	Completed
Columbia Generating Station*	USA	1984	Completed
Comanche Peak Steam Electric Station	USA	1990	Completed

Name of Facility	Country	Year of Decision	Outcome
Connecticut Yankee Nuclear Power Plant*	USA	1968	Completed
Cooper Nuclear Station	USA	1974	Completed
Crystal River Nuclear Generating Plant	USA	1977	Completed
Davis-Besse Nuclear Power Station	USA	1978	Completed
Diablo Canyon Nuclear Power Plant*	USA	1985	Completed
Donald C. Cook Nuclear Power Plant	USA	1975	Completed
Douglas Point Nuclear Generating Station*	USA	1977	Cancelled
Dresden Nuclear Power Station	USA	1960	Completed
Duane Arnold Energy Center	USA	1974	Completed
Edwin I. Hatch Nuclear Plant	USA	1974	Completed
Enrico Fermi Atomic Power Plant	USA	1972	Completed
Erie Nuclear Power Plant*	USA	1980	Cancelled
Forked River Nuclear Power Plant	USA	1980	Cancelled
Fort Calhoun Station	USA	1973	Completed
Fort St. Vrain Generating Station	USA	1977	Completed
Grand Gulf Nuclear Station	USA	1985	Completed
Greene County Nuclear Power Plant*	USA	1979	Cancelled
H. B. Robinson Steam Electric Plant*	USA	1971	Completed
Hartsville Nuclear Plant*	USA	1984	Cancelled
Haven Nuclear Power Plant*	USA	1980	Cancelled
Hope Creek Generating Station	USA	1986	Completed
Humboldt Bay Nuclear Power Plant	USA	1963	Completed
Indian Point Nuclear Generating*	USA	1962	Completed
James A. FitzPatrick Nuclear Power Plant	USA	1975	Completed
Joseph M. Farley Nuclear Plant	USA	1977	Completed
Kewaunee Power Station	USA	1974	Completed

(Continued)

Name of Facility	Country	Year of Decision	Outcome
LaSalle County Station	USA	1982	Completed
Limerick Generating Station*	USA	1986	Completed
Lyons Kansas Nuclear Waste Repository	USA	1972	Cancelled
Maine Yankee Nuclear Power Plant*	USA	1972	Completed
Marble Hill Nuclear Power Plant	USA	1985	Cancelled
McGuire Nuclear Station	USA	1981	Completed
Midland Cogeneration Venture	USA	1986	Cancelled
Millstone Power Station	USA	1970	Completed
Montague Nuclear Power Plant	USA	1980	Cancelled
Monticello Nuclear Generating Plant	USA	1971	Completed
Nine Mile Point Nuclear Station	USA	1969	Completed
North Anna Power Station	USA	1978	Completed
Oconee Nuclear Station	USA	1973	Completed
Oyster Creek Nuclear Generating Station	USA	1969	Completed
Palisades Nuclear Plant*	USA	1971	Completed
Palo Verde Nuclear Generating Station	USA	1988	Completed
Pathfinder Atomic Power Plant	USA	1966	Completed
Peach Bottom Atomic Power Station	USA	1966	Completed
Pebble Springs Nuclear Power Plant	USA	1982	Cancelled
Perkins Nuclear Power Plant	USA	1982	Cancelled
Perry Nuclear Power Plant*	USA	1986	Completed
Phipps Bend Nuclear Power Plant	USA	1982	Cancelled
Pilgrim Nuclear Power Station	USA	1972	Completed
Point Beach Nuclear Plant	USA	1970	Completed
Prairie Island Nuclear Generating Plant	USA	1973	Completed
Quad Cities Nuclear Power Station	USA	1972	Completed

Name of Facility	**Country**	**Year of Decision**	**Outcome**
Quanicassee Nuclear Power Plant	USA	1974	Cancelled
R.E. Ginna Nuclear Power Plant	USA	1970	Completed
Rancho Seco Nuclear Generating Station	USA	1975	Completed
River Bend Station	USA	1986	Completed
Salem Nuclear Generating Station	USA	1977	Completed
San Onofre Nuclear Generating Station	USA	1968	Completed
Saxton Nuclear Generating Station	USA	1961	Completed
Seabrook Station	USA	1990	Completed
Sears Isle Nuclear Power Plant	USA	1977	Cancelled
Sequoyah Nuclear Plant*	USA	1981	Completed
Shearon Harris Nuclear Power Plant	USA	1987	Completed
Shippingport Atomic Power Station*	USA	1957	Completed
Shoreham Nuclear Power Plant*	USA	1989	Cancelled
Skagit Nuclear Power Plant	USA	1983	Cancelled
Skull Valley Repository	USA	2007	Cancelled
Somerset Nuclear Power Plant/Kintigh Generating Station	USA	1975	Cancelled
South River Nuclear Power Plant*	USA	1978	Cancelled
South Texas Project*	USA	1988	Completed
St. Lucie Plant*	USA	1976	Completed
Stanislaus Nuclear Power Plant	USA	1979	Cancelled
Sterling Nuclear Plant	USA	1980	Cancelled
Sundesert Nuclear Power Plant*	USA	1978	Cancelled
Surry Nuclear Power Station	USA	1972	Completed
Susquehanna Steam Electric Station	USA	1982	Completed
Three Mile Island Nuclear Station	USA	1974	Completed
Trojan Nuclear Power Plant	USA	1976	Completed

(Continued)

Name of Facility	Country	Year of Decision	Outcome
Turkey Point Nuclear Generating	USA	1972	Completed
Tyrone Nuclear Power Plant	USA	1979	Cancelled
Vandalia Nuclear Project	USA	1982	Cancelled
Vermont Yankee Nuclear Power Plant	USA	1973	Completed
Virgil C. Summer Nuclear Station	USA	1984	Completed
Vogtle Electric Generating Plant	USA	1987	Completed
Waste Isolation Pilot Plant	USA	1999	Completed
Waterford Steam Electric Station	USA	1985	Completed
Watts Bar Nuclear Plant	USA	1996	Completed
William H. Zimmer Power Station	USA	1984	Cancelled
Wolf Creek Generating Station	USA	1985	Completed
Yankee Rowe Nuclear Power Station	USA	1960	Completed
Yellow Creek Nuclear Power Plant	USA	1984	Cancelled
Zion Nuclear Power Station	USA	1973	Completed

(* Indicates case was included in the randomly drawn subset of cases for Stage 2 analysis)

A.2. The Polity IV Project

The Polity IV Project is centered around the tradition of coding the authority characteristics of states in the world system for purposes of comparative, quantitative analysis. The original Polity conceptual scheme was formulated, and the original Polity I data collected, under the direction of Ted Robert Gurr; the Polity scheme was informed by foundational, collaborative work with Harry Eckstein, *Patterns of Authority: A Structural Basis for Political Inquiry* (New York: John Wiley & Sons 1975). The Polity project has proven its value to researchers over the years, becoming the most widely used data resource for studying regime change and the effects of regime authority.

The Polity IV Project carries data collection and analysis through 2010 and is under the direction of Dr. Monty G. Marshall and supported by the Political Instability Task Force, Societal-Systems Research, and Center for Systemic Peace.

The Polity conceptual scheme is unique in that it examines *concomitant qualities of democratic and autocratic authority* in governing institutions, rather than discreet and mutually exclusive forms of governance. This perspective envisions a spectrum of governing authority that spans from *fully institutionalized autocracies* through *mixed, or incoherent, authority regimes* (termed "anocracies") to *fully institutionalized democracies*. The "Polity Score" captures this regime authority spectrum on a 21- point scale ranging from -10 (hereditary monarchy) to +10 (consolidated democracy). The Polity scores can also be converted to regime categories: we recommend a three-part categorization of "autocracies" (- 10 to -6), "anocracies" (-5 to +5 and the three special values: -66, -77, and -88), and "democracies" (+6 to +10); see "Global Regimes by Type, 1946-2006" above. The Polity scheme consists of six component measures that record key qualities of executive recruitment, constraints on executive authority, and political competition. It also records changes in the institutionalized qualities of governing authority. The Polity data include information only on the institutions of the central government and on political groups acting, or reacting, within the scope of that authority. It does not include consideration of groups and territories that are actively removed from that authority (i.e., separatists or "fragments;" these are considered separate, though not independent, polities) or segments of the population that are not yet effectively politicized in relation to central state politics.

A.3. Case Material References

Common Sources

Quality of Democracy

Center for Systemic Peace: Integrated Network for Societal Conflict Research. 2010. "Polity IV Annual Time-Series 1800-2011." http://www.systemicpeace.org/inscr/inscr.htm. Last accessed on October 24, 2012.

Decision Year

International Atomic Energy Agency. 2012. "Power Reactor Information System." http://www.iaea.org/PRIS/home.aspx. Last accessed on October 24, 2012.

Case Specific Sources

Armenia

Metsamor Nuclear Power Plant

Federalism

Department Of State. The Office of Website Management, Bureau of Public Affairs. 2012. "Armenia." *U.S. Department of State*. http://www.state.gov/r/pa/ei/bgn/5275.htm. *Last accessed on October 15, 2012.*

Soviet Union Information Bureau. 1929. "The Soviet Union: Facts, Descriptions, Statistics — Ch 2."http://www.marxists.org/history/ussr/government/1928/sufds/ch02.htm. Last accessed on October 15, 2012.

Brazil

Angra Nuclear Power Plant

Federalism

Forum of Federations. "Federalism in Brazil." http://www.forumfed.org/en/where/brazil.php. Last accessed on October 15, 2012.

Bulgaria

Kozloduy Nuclear Power Plant

Federalism

Ganev, Venelin. 2009. "Bulgaria Under Communism: 1944-1989." *National Exhibit: Bulgaria Exhibit*. http://bulgaria. Globalmuseumon communism.org/bulgaria/history. Last accessed on October 15, 2012.

Czechoslovakia

Bohunice Nuclear Power Plant

Federalism

Alfred C. Stepan. 1999. "Federalism and Democracy: Beyond the U.S. Model." *Journal of Democracy* 10(4): 19–34.

France

Saint Alban Nuclear Power Plant

Federalism

Democracy Web: Comparative Studies in Freedom. "Constitutional Limits on Government: France." http://www.democracyweb.org/limits/france.php. Last accessed on October 15, 2012.

Saint Laurent Nuclear Power Plant

Activism

Aldrich, Daniel P. 2008. *Site Fights: Divisive Facilities and Civil Society in Japan and the West.* Cornell University Press, p. 167.

Federalism

Democracy Web: Comparative Studies in Freedom. "Constitutional Limits on Government: France." http://www.democracyweb.org/limits/france.php. Last accessed on October 15, 2012.

Public Engagement

Aldrich, Daniel P. 2008. *Site Fights: Divisive Facilities and Civil Society in Japan and the West.* Cornell University Press, p. 167.

Golfech Nuclear Power Plant

Activism

Aldrich, Daniel P. 2008. *Site Fights: Divisive Facilities and Civil Society in Japan and the West.* Cornell University Press, pp. 165-166.

Federalism

Democracy Web: Comparative Studies in Freedom. "Constitutional Limits on Government: France." http://www.democracyweb.org/limits/france.php. Last accessed on October 15, 2012.

Public Engagement

Aldrich, Daniel P. 2008. *Site Fights: Divisive Facilities and Civil Society in Japan and the West.* Cornell University Press, pp. 165-166.

Blayais Nuclear Power Plant

Activism

Aldrich, Daniel P. 2008. *Site Fights: Divisive Facilities and Civil Society in Japan and the West.* Cornell University Press, p. 181.

Federalism

Democracy Web: Comparative Studies in Freedom. "Constitutional Limits on Government: France." Fhttp://www.democracyweb.org/limits/france.php. Last accessed on October 15, 2012.

Flamanville Nuclear Power Plant

Activism

Aldrich, Daniel P. 2008. *Site Fights: Divisive Facilities and Civil Society in Japan and the West.* Cornell University Press, p. 167.

Federalism

Democracy Web: Comparative Studies in Freedom. "Constitutional Limits on Government: France." Fhttp://www.democracyweb.org/limits/france.php. Last accessed on October 15, 2012.

Public Engagement

Aldrich, Daniel P. 2008. *Site Fights: Divisive Facilities and Civil Society in Japan and the West.* Cornell University Press, p. 167.

Germany

Greifswald Nuclear Power Station

Federalism

Alfred C. Stepan. 1999. "Federalism and Democracy: Beyond the U.S. Model." *Journal of Democracy*10(4): 19–34.

SNR-300 Fast Breeder Reactor

Activism

"Wunderland Kalkar: Nuclear Power Plant Turned Amusement Park." 2011. *Amusing Planet*. http://www.amusingplanet.com/2011/06/wunderland-kalkar-nuclear-power-plant.html. Last accessed on October 10, 2012.

Federalism

"Germany: Government." 2012. *CIA - The World Factbook*. https://www.cia.gov/library/publications/the- world-factbook/geos/gm. html. Last accessed on October 15, 2012.

Public Engagement

Marth, Willy. 1994. "The SNR-300 Fast Breeder in the Ups and Downs of Its History." http://bibliothek.fzk.de/zb/kfk-berichte/KFK5455.pdf.

Kruemmel Nuclear Power Plant

Federalism

"Germany: Government." 2012. *CIA - The World Factbook*. https://www.cia.gov/library/publications/the- world-factbook/geos/gm. html. Last accessed on October 15, 2012.

India

Rajasthan Atomic Power Station

Federalism

Praveen Gupta. "Federalism: New Dimensions in Indian Context." *Lawyers Club India*. http://www.lawyersclubindia.com/articles/Federalism-New-Dimensions-in-Indian-Context- 3480.asp. Last accessed on October 15, 2012.

Italy

Montalto di Castro (Alto Lazio)

Activism

Giugni, Marco. 2004. *Social Protest and Policy Change: Ecology, Antinuclear, and Peace Movements in Comparative Perspective*. Rowman & Littlefield, p. 54.

Jenkins, J. C., & Klandermans, B. (Eds.). 1995. *The politics of social protest: Comparative perspectives on states and social movements* (Vol. 3). University of Minnesota Press, pp. 240-241.

World Nuclear Association. 2012. "Nuclear Power in Italy." *World Nuclear Association.* http://www.world-nuclear.org/info/inf101.html. Last accessed October 4, 2012.

Federalism

Caravita, Beniamino. 2004. "Italy Towards a Federal State?" http://camlaw.rutgers.edu/statecon/subpapers/caravita.pdf. Last accessed October 17, 2012.

Public Engagement

World Nuclear Association. 2012. "Nuclear Power In Italy." *World Nuclear Association.* http://www.world-nuclear.org/info/inf101.html. Last accessed on October 10, 2012.

Enrico Fermi (Trino Vercellese)

Activism

Giugni, Marco. 2004. *Social Protest and Policy Change: Ecology, Antinuclear, and Peace Movements in Comparative Perspective.* Rowman & Littlefield, p. 54.

Federalism

Caravita, Beniamino. 2004. "Italy Towards a Federal State?" http://camlaw.rutgers.edu/statecon/subpapers/caravita.pdf. Last accessed on October 17, 2012.

Japan

Kashiwazaki-Kariwa Nuclear Power Plant

Activism

Aldrich, Daniel P. 2008. *Site Fights: Divisive Facilities and Civil Society in Japan and the West.* Cornell University Press, p. 129-131.

Surrey, John, and Charlotte Huggett. 1976. "Opposition to nuclear power: A review of international experience." *Energy Policy* 4(4): 286–307.

Federalism

Greenleaf, Graham. 2010. *Comparative Study on Different Approaches to New Privacy Challenges, In Particular in the Light of Technological Developments.* European Commission. http://ec.europa.eu/justice/policies/privacy/docs/studies/new_privacy_challenges/final_report_cou ntry_report_B5_japan.pdf. Last accessed on October 17, 2012.

Public Engagement

Aldrich, Daniel P. 2008. *Site Fights: Divisive Facilities and Civil Society in Japan and the West*. Cornell University Press, p. 129-131.

Maki Nuclear Power Plant

Activism

Aldrich, Daniel P. 2008. *Site Fights: Divisive Facilities and Civil Society in Japan and the West*. Cornell University Press, p. 139.

Federalism

Greenleaf, Graham. 2010. *Comparative Study on Different Approaches to New Privacy Challenges, In Particular in the Light of Technological Developments*. European Commission. http://ec.europa.eu/justice/policies/privacy/docs/studies/new_privacy_challenges/final_report_country_report_B5_japan.pdf. Last accessed on October 17, 2012.

Public Engagement

Aldrich, Daniel P. 2008. *Site Fights: Divisive Facilities and Civil Society in Japan and the West*. Cornell University Press, p. 139.

Tomari Nuclear Power Plant

Activism

Aldrich, Daniel P. 2008. *Site Fights: Divisive Facilities and Civil Society in Japan and the West*. Cornell University Press, p. 129.

Federalism

Greenleaf, Graham. 2010. *Comparative Study on Different Approaches to New Privacy Challenges, In Particular in the Light of Technological Developments*. European Commission. http://ec.europa.eu/justice/policies/privacy/docs/studies/new_privacy_challenges/final_report_country_report_B5_japan.pdf. Last accessed on October 17, 2012.

Public Engagement

Aldrich, Daniel P. 2008. *Site Fights: Divisive Facilities and Civil Society in Japan and the West*. Cornell University Press, p. 129.

Netherlands

Borssele Nuclear Power Plant

Federalism

Toonen. 2007. "The Netherlands: A Decentralised Unitary State in a Welfare Society." *West European Politics* 10(4): 108–29.

South Korea

Wolseong Nuclear Power Plant

Federalism

Chong-Min, Park. 2006. "Local Governance and Community Power in Korea." *Korea Journal* 46(4): 9– 32.

"Korean Constitution." 1987. *Constitutional Court of Korea*. http://english.ccourt.go.kr/home/att_file/download/Constitution_of_the_Republic_of_Korea.pdf. Last accessed on October 17, 2012.

Uljin Nuclear Power Plant

Federalism

Chong-Min, Park. 2006. "Local Governance and Community Power in Korea." *Korea Journal* 46(4): 9– 32.

"Korean Constitution." 1987. *Constitutional Court of Korea*. http://english.ccourt.go.kr/home/att_file/download/Constitution_of_the_Republic_of_Korea.pdf. Last accessed on October 17, 2012.

Spain

Cofrentes Nuclear Power Plant

Federalism

Conversi, Daniele. 1998. "The Spanish Federalist Tradition and the 1978 Constitution." *telos* 1998(112): 125–44.

Pi-Sunyer, Carles Viver. 2010. *The Transition to a Decentralized Political System in Spain*. Forum of Federations. http://www.forumfed.org/pubs/occasional_papers/OCP4.pdf. Last accessed on October 17, 2012.

Sayago

Federalism

Conversi, Daniele. 1998. "The Spanish Federalist Tradition and the 1978 Constitution." *telos* 1998(112): 125–44.

Pi-Sunyer, Carles Viver. 2010. *The Transition to a Decentralized Political System in Spain*. Forum of Federations. http://www.forumfed.org/pubs/occasional_papers/OCP4.pdf. Last accessed on October 17, 2012.

Sweden

Oskarshamn Nuclear Power Plant

Federalism

Erlingsson, Gissur O, and Jorgen Odalen. "How Should Local Government Be Organized? Reflections from a Swedish Perspective." http://student.statsvet.uu.se/modules/kurser/visadokument.aspx?id=13527. Last accessed on October 17, 2012.

Switzerland

Gosgen Nuclear Power Plant

Activism

Giugni, Marco. 2004. *Social Protest and Policy Change: Ecology, Antinuclear, and Peace Movements in Comparative Perspective*. Rowman & Littlefield Publishers, INC., p. 65.

Gyorgy, Anna. 1979. *No Nukes: Everyone's Guide to Nuclear Power*. Black Rose Books Ltd., p. 324.

Federalism

"Federalism Is a Precious Principle to the Swiss." 2009. *swissinfo.ch*. http://www.swissinfo.ch/eng/politics/Precious_principle_of_federalism.html?cid=37186. Last accessed on October 29, 2012.

Public Engagement

International Atomic Energy Association. 2010. "Switzerland." *International Atomic Energy Association*. http://www.pub.iaea.org/MTCD/publications/PDF/CNPP2010_CD/countryprofiles/Switzerland/CNPP2010Switzerland.htm. Last accessed on October 15, 2012.

United Kingdom

Hartlepool Nuclear Power Station

Federalism

Goodlad, Robina. "Political Lobbying and the Unitary State: The Case of Scottish Homelessness Legislation." *Scottish Affairs* 2000(30). http://www.scottishaffairs.org/backiss/pdfs/sa30/SA30_Goodlad.pdf. Last accessed on October 29, 2012.

Hinkley Point Nuclear Power Station

Federalism

Goodlad, Robina. "Political Lobbying and the Unitary State: The Case of Scottish Homelessness Legislation." *Scottish Affairs* 2000(30). http://www.scottishaffairs.org/backiss/pdfs/sa30/SA30_Goodlad.pdf. Last accessed on October 29, 2012.

Hunterston Nuclear Power Station

Federalism

Goodlad, Robina. "Political Lobbying and the Unitary State: The Case of Scottish Homelessness Legislation." *Scottish Affairs* 2000(30). http://www.scottishaffairs.org/backiss/pdfs/sa30/SA30_Goodlad.pdf. Last accessed on October 29, 2012.

Torness Nuclear Power Station

Activism

Surrey, John, and Charlotte Huggett. 1976. "Opposition to nuclear power: A review of international experience." *Energy Policy* 4(4): 286–307.

Williams, Roger. 1980. *The Nuclear Power Decisions: British Policies, 1953-78*. Croom Helm, p. 261.

Federalism

Goodlad, Robina. "Political Lobbying and the Unitary State: The Case of Scottish Homelessness Legislation." *Scottish Affairs* 2000(30). http://www.scottishaffairs.org/backiss/pdfs/sa30/SA30_Goodlad.pdf. Last accessed on October 29, 2012.

United States

Atlantic Nuclear Power Plant

Activism

Janson, Donald. 1978. "Report Cites Peril in Mishap at Offshore Atom Plant; Much 'Prodding' Reported Opposition During Campaign." *The New York Times*: p. 27.

McPhee, John. 1975. "A Reporter at Large: The Atlantic Nuclear Generating Station." *The New Yorker*: p. 98.

Federalism

Stepan, Alfred C. 1999. "Federalism and Democracy: Beyond the U.S. Model." *Journal of Democracy* 10(4): 19–34.

Public Engagement

McPhee, John. 1975. "A Reporter at Large: The Atlantic Nuclear Generating Station." *The New Yorker*: p. 98.

Beaver Valley Power Station

Activism

Sneddon, James. 1976b. "Plant Start-Up Date Uncertain." *Beaver County Times*: 18.

"Six Anti-nuclear Protesters Jailed After Beaver Falls Incident." 1979. *Lakeland Ledger*: p. 4

Federalism

Stepan, Alfred C. 1999. "Federalism and Democracy: Beyond the U.S. Model." *Journal of Democracy* 10(4): 19–34.

Public Engagement

Sneddon, James. 1976a. "Hearing Process Nears End." *Beaver County Times*: 18. Sneddon, James. 1976b. "Plant Start-Up Date Uncertain." *Beaver County Times*: 18.

Sneddon, James. 1975. "Dusquesne Light has one more hearing to hurdle." *Beaver County Times*.

Carroll County Nuclear Power Plant

Federalism

Stepan, Alfred C. 1999. "Federalism and Democracy: Beyond the U.S. Model." *Journal of Democracy* 10(4): 19–34.

Columbia Generating Station

Federalism

Stepan, Alfred C. 1999. "Federalism and Democracy: Beyond the U.S. Model." *Journal of Democracy* 10(4): 19–34.

Public Engagement

"Money running short for Hanford nuclear plant." 1982. *Lewiston Morning Tribune*.

Connecticut Yankee Nuclear Power Station

Federalism

Stepan, Alfred C. 1999. "Federalism and Democracy: Beyond the U.S. Model." *Journal of Democracy* 10(4): 19–34.

Diablo Canyon Nuclear Power Plant

Activism

"450 Protestors Arrested at Nuclear Power Plant." 1981. *The Bryan Times*. http://news.google.com/newspapers?id=p7BPAAAAIBAJ&sjid=SFIDAAAAIBAJ&pg=4745,7046152&dq=diablo-canyon+nuclear+power+plant&hl=en. Last accessed on September 26, 2012.

Federalism

Stepan, Alfred C. 1999. "Federalism and Democracy: Beyond the U.S. Model." *Journal of Democracy* 10(4): 19–34.

Public Engagement

"Number of Demonstrators at Diablo Canyon Is Dwindling." 1981. *Merced Sun-Star*. http://news.google.com/newspapers?id=P7JcAAAAIBAJ&sjid=o1gNAAAAIBAJ&pg=7048,3858103&dq=diablo-canyon+nuclear+power+plant+hearings&hl=en. Last accessed on September 26, 2012.

Douglas Point Nuclear Generating Station

Activism

Epstein, Daniel. 1976. "Maryland Nuclear Plant Faces Fight at Hearings*."* *The Free Lance-Star*. http://news.google.com/newspapers?id= c95NAAAAIBAJ&sjid=9IoDAAAAIBAJ&pg=1268,2448833&dq=douglas+point+nuclear+generating+station+maryland&hl=en. Last accessed on September 26, 2012.

Sullivan, Paul. 1976. "Nuclear Power Plant Hearings Off to Quite Start." *The Free Lance-Star*. http://news.google.com/newspapers?id= d95NAAAAIBAJ&sjid=9IoDAAAAIBAJ&pg=5698,1660963&dq=douglas-point+nuclear+generating+station&hl=en. Last accessed on September 26, 2012.

Federalism

Stepan, Alfred C. 1999. "Federalism and Democracy: Beyond the U.S. Model." *Journal of Democracy* 10(4): 19–34.

Public Engagement

Epstein, Daniel. 1976. "Maryland Nuclear Plant Faces Fight at Hearings." *The Free Lance-Star*. http://news.google.com/newspapers?id= c95NAAAAIBAJ&sjid=9IoDAAAAIBAJ&pg=1268,2448833&dq=douglas+point+nuclear+generating+station+maryland&hl=en. Last accessed on September 26, 2012.

Sullivan, Paul. 1976. "Nuclear Power Plant Hearings Off to Quite Start." *The Free Lance-Star*. http://news.google.com/newspapers?id= d95NAAAAIBAJ&sjid=9IoDAAAAIBAJ&pg=5698,1660963&dq=douglas-point+nuclear+generating+station&hl=en. Last accessed on September 26, 2012.

Erie Nuclear Power Plant

Activism

"Radiation, Cancer Linked." 1978. *The News-Dispatch*. http://news.google.com/newspapers?id=rExVAAAAIBAJ&sjid=cT4NAAAAIBAJ&pg=5147,3826183&dq=erie+nuclear+power+plant&hl=en. Last accessed on September 26, 2012.

"Erie power plant opposed." 1977. *Ludington Daily News*. http://news.google.com/newspapers?nid=110&dat=19771027&id=sR8vAAAAIBAJ&sjid=y9wF AAAAIBAJ&pg=6263,2626145. Last accessed on September 26, 2012.

Federalism

Stepan, Alfred C. 1999. "Federalism and Democracy: Beyond the U.S. Model." *Journal of Democracy* 10(4): 19–34.

Public Engagement

"Radiation, Cancer Linked." 1978. *The News-Dispatch*. http://news.google.com/newspapers?id=rExVAAAAIBAJ&sjid=cT4NAAAAIBAJ&pg=5147,38 26183&dq=erie+nuclear+power+plant&hl=en. Last accessed on September 26, 2012.

Greene County Nuclear Power Plant

Federalism

Stepan, Alfred C. 1999. "Federalism and Democracy: Beyond the U.S. Model." *Journal of Democracy* 10(4): 19–34.

H.B. Robinson Steam Electric Plant

Federalism

Stepan, Alfred C. 1999. “Federalism and Democracy: Beyond the U.S. Model.” *Journal of Democracy* 10(4): 19–34.

Hartsville Nuclear Plant

Activism

“Protests Growing Use of Nuclear Power: Catfish Alliance Releases Balloons.” 1977. *The News and Courier.* http://news.google.com/newspapers?id=b_s0AAAAIBAJ&sjid=BE8KAAAAIBAJ&pg=1145,1160068&dq=hartsville+nuclear+plant+protest&hl=en. Last accessed on September 27, 2012.

“Nuclear hearing asked.” 1977. *Kentucky New Era.* http://news.google.com/newspapers?id=x5c0AAAAIBAJ&sjid=w2gFAAAAIBAJ&pg=3652,740496&dq=hartsville+nuclear+plant&hl=en. Last accessed on September 26, 2012.

Federalism

Stepan, Alfred C. 1999. “Federalism and Democracy: Beyond the U.S. Model.” *Journal of Democracy* 10(4): 19–34.

Public Engagement

“TVA Accused of Doctoring Future Power Demand Figures.” 1975. *Times Daily.* http://news.google.com/newspapers?id=CQIsAAAAIBAJ&sjid=LMgEAAAAIBAJ&pg=2865,4735423&dq=hartsville+nuclear+plant+public+hearings&hl=en. Last accessed on September 27, 2012.

“Nuclear plant opposed.” 1977. *Lakeland Ledger.* http://news.google.com/newspapers?id=TnVhAAAAIBAJ&sjid=-voDAAAAIBAJ&pg=3741,2026270&dq=hartsville+nuclear+plant&hl=en. Last accessed on September 26, 2012.

Haven Nuclear Power Plant

Activism

“Nuclear loan bid rejected at bank firm.” 1980. *The Milwaukee Sentinel.* http://news.google.com/newspapers?id=MnpQAAAAIBAJ&sjid=AxIEAAAAIBAJ&pg=6547,3099827&dq=haven-nuclear-plant&hl=en. Last accessed September 26, 2012.

Federalism

Stepan, Alfred C. 1999. “Federalism and Democracy: Beyond the U.S. Model.” *Journal of Democracy* 10(4): 19–34.

Public Engagement

“Sound Decision to Drop Haven.” 1980. *The Milwaukee Journal.* http://news.google.com/newspapers?id=lF0aAAAAIBAJ&sjid=-CoEAAAAIBAJ&pg=6365,1505359&dq=haven+nuclear+power+plant&hl=en. Last accessed on September 27, 2012.

Indian Point Nuclear Generating Station

Federalism
Stepan, Alfred C. 1999. "Federalism and Democracy: Beyond the U.S. Model." *Journal of Democracy*10(4): 19–34.
Public Engagement
"Indian Point Timeline... a Troubling History." *Indian Point Safe Energy Coalition*. http://www.ipsecinfo.org/1952.htm. Last accessed on September 27, 2012.
Limerick Generating Station
Activism
Smith, Erich. 1983. "Delaware River Water Diversion Project Causing Fight." *Gettysburg Times*. http://news.google.com/newspapers?id= qKgyAAAAIBAJ&sjid=3OcFAAAAIBAJ&pg=1249,530131&dq=limerick+generating+station+protests&hl=en. Last accessed on September 27, 2012.
"Nuke plant funding is denied." 1982. *Youngstown Vindicator*. http://news.google.com/newspapers?id=fidAAAAAIBAJ&sjid=qVgMAAAAIBAJ&pg=3053,2883697&dq=limerick-nuclear&hl=en. Last accessed on September 27, 2012.
Federalism
Stepan, Alfred C. 1999. "Federalism and Democracy: Beyond the U.S. Model." *Journal of Democracy*10(4): 19–34.
Public Engagement
Barker. 1985. "Ruling Opens Way for Power License at Limerick Plant." *Pittsburgh Post-Gazette*. http://news.google.com/newspapers?id= boVIAAAAIBAJ&sjid=i24DAAAAIBAJ&pg=6747,5496072&dq=limerick+generating+station+public&hl=en. Last accessed on September 27, 2012.
"Calling total halt to Seabrook Nuclear Power Plant." 1981. *The Lewiston Daily Sun*. http://news.google.com/newspapers?id= MIAjAAAAIBAJ&sjid=i2cFAAAAIBAJ&pg=1364,1137123&dq=limerick-nuclear&hl=en. Last accessed September 27, 2012.
Maine Yankee Nuclear Power Plant
Activism
"Maine May Opt Out of Nuclear Era." 1980. *Eugene Register-Guard*. http://news.google.com/newspapers?id=R_JVAAAAIBAJ&sjid=6eEDAAAAIBAJ&pg=6783,3156823&dq=maine-yankee-nuclear-power-plant+opposition&hl=en. Last accessed on September27, 2012.
Federalism
Stepan, Alfred C. 1999. "Federalism and Democracy: Beyond the U.S. Model." *Journal of Democracy*10(4): 19–34.

Public Engagement

"Maine Yankee Hearing Will Begin on Wednesday." 1972. *Bangor Daily News.* http://news.google.com/newspapers?id=0ic0AAAAIBAJ&sjid=5eAIAAAAIBAJ&pg=676,1354257&dq=maine+yankee+nuclear+power+plant&hl=en. Last accessed on September 27, 2012.

Palisades Nuclear Power Plant

Activism

Kilpatrick, James. 1972. "The High Price of Clean Air." *Sarasota Journal.* http://news.google.com/newspapers?id=aXU0AAAAIBAJ&sjid=GY0EAAAAIBAJ&pg=7533,5 376121&dq=palisades+nuclear+plant&hl=en. Last accessed on September 27, 2012.

"Provisional license granted to consumers for operation of Palisades Nuclear Plant." 1971. *Ludington Daily News.* http://news.google.com/newspapers?id=crpNAAAAIBAJ&sjid=FkoDAAAAIBAJ&pg=4539,4669470&dq=palisades+nuclear+plant&hl=en. Last accessed on September 27, 2012.

Federalism

Stepan, Alfred C. 1999. "Federalism and Democracy: Beyond the U.S. Model." *Journal of Democracy* 10(4): 19–34.

Public Engagement

"Hearings Open on Effects of Consumers' New Plant." 1970. *The Owosso Argus-Press.* http://news.google.com/newspapers?id=oTUiAAAAIBAJ&sjid=6KoFAAAAIBAJ&pg=4247,5518052&dq=palisades+nuclear+plant+public+hearings&hl=en. Last accessed on September 27, 2012.

"Nuclear plant opening will be delayed." 1971. *The Owosso Argus-Press.* http://news.google.com/newspapers?id=hVwiAAAAIBAJ&sjid=iqsFAAAAIBAJ&pg=760,3050 954&dq=palisades+nuclear+plant&hl=en. Last accessed on September 27, 2012.

Perry Nuclear Power Plant

Activism

"Hundreds Are Arrested in Anti-nuclear Protests." *Toledo Blade.* 1979. http://news.google.com/newspapers?id=nUlPAAAAIBAJ&sjid=OQMEAAAAIBAJ&pg=7086,7 665485&dq=perry+nuclear+power+plant&hl=en. Last accessed on September 27, 2012.

Spetz, John. 1979. "Opponents March Against Perry Nuclear Plant." *The Bryan Times.* http://news.google.com/newspapers?id=6L8wAAAAIBAJ&sjid=x1EDAAAAIBAJ&pg=4893,6343356&dq=perry+nuclear+generating+station&hl=en. Last accessed on October 1, 2012.

Federalism

Stepan, Alfred C. 1999. "Federalism and Democracy: Beyond the U.S. Model." *Journal of Democracy*10(4): 19–34.

Public Engagement

"Questions Still Dog Perry Plant but Officials Tell Congress Facility Met Test After January's Quake." 1986. *The Vindicator*. http://news.google.com/newspapers?id=_l1cAAAAIBAJ&sjid=fVYNAAAAIBAJ&pg=3946,215000&dq=perry+nuclear+generating+station+hearings&hl=en. Last accessed on October 1, 2012.

"Perry Nuclear Plant advances." 1985. *Toledo Blade*. http://news.google.com/newspapers?id=YTJPAAAAIBAJ&sjid=zgIEAAAAIBAJ&pg=3803,1459532&dq=perry-nuclear&hl=en. Last accessed on September 27, 2012.

Sequoyah Nuclear Plant

Activism

"200 Protest At Plant in Tenn." 1980. *Herald-Journal*. http://news.google.com/newspapers?id=I08sAAAAIBAJ&sjid=sM0EAAAAIBAJ&pg=2875,6412015&dq=sequoyah+nuclear+plant&hl=en. Last accessed on October 1, 2012.

"Nuclear plant fueled up." 1980. *The Morning Record and Journal*. http://news.google.com/newspapers?id=owlIAAAAIBAJ&sjid=cQANAAAAIBAJ&pg=2189,12 6432&dq=sequoyah-nuclear&hl=en. Last accessed on September 27, 2012.

"Operators at Sequoyah work on turbine trouble at plant." 1980. *Kentucky New Era*. http://news.google.com/newspapers?id=Rx4vAAAAIBAJ&sjid=0NwFAAAAIBAJ&pg=2860,36 67157&dq=sequoyah-nuclear&hl=en. Last accessed on September 27, 2012.

Federalism

Stepan, Alfred C. 1999. "Federalism and Democracy: Beyond the U.S. Model." *Journal of Democracy*10(4): 19–34.

Public Engagement

"Around the Nation; Pills to Be Distributed In Nuclear Plant Area." *The New York Times*. 1981. http://www.nytimes.com/1981/11/03/us/around-the-nation-pills-to-be-distributed-in-nuclear-plant-area.html. Last accessed on October 1, 2012.

"Residents not worried about nuclear accident." 1981. *The Times-News*. http://news.google.com/newspapers?id=KiAaAAAAIBAJ&sjid=LSQEAAAAIBAJ&pg=6912,23 44296&dq=sequoyah-nuclear&hl=en. Last accessed on September 27, 2012.

Shippingport Atomic Power Station

Activism

"A nuclear drama bring enacted at Shippingport." 1979. *The Albany Sunday Herald.* http://news.google.com/newspapers?id=tmpEAAAAIBAJ&sjid=ZLIMAAAAIBAJ&pg=4539,3548031&dq=beaver-valley-nuclear+public&hl=en. Last accessed on September 24, 2012.

Federalism

Stepan, Alfred C. 1999. "Federalism and Democracy: Beyond the U.S. Model." *Journal of Democracy* 10(4): 19–34.

Shoreham Nuclear Power Plant

Activism

"Nuclear Demonstrators Arrested but No Violence at Power Plant." 1980. *Lakeland Ledger.* http://news.google.com/newspapers?id= nkBIAAAAIBAJ&sjid=IvsDAAAAIBAJ&pg=6996,4837649&dq=shoreham+nuclear+power+plant&hl=en. Last accessed on October 1, 2012.

"147 nuclear protestors held." 1980. *The Vancouver Sun.* http://news.google.com/newspapers?id=h5tlAAAAIBAJ&sjid=HYwNAAAAIBAJ&pg=4311,4457475&dq=shoreham-nuclear&hl=en. Last accessed on September 27, 2012.

"Shoreham Plant's emergency drill to be tested today, despite foes." 1986. *Schenectady Gazette.* http://news.google.com/newspapers?id=TkJGAAAAIBAJ&sjid=_OcMAAAAIBAJ&pg=1214,2501031&dq=shoreham-nuclear+hearing&hl=en. Last accessed on September 27, 2012.

Federalism

Stepan, Alfred C. 1999. "Federalism and Democracy: Beyond the U.S. Model." *Journal of Democracy*10(4): 19–34.

South River Nuclear Power Plant

Federalism

Stepan, Alfred C. 1999. "Federalism and Democracy: Beyond the U.S. Model." *Journal of Democracy*10(4): 19–34.

South Texas Project

Activism

United Press International. 1979. "500 Arrested at Anti-nuclear Protest." *The Telegraph-Herald.* http://news.google.com/newspapers?id= 2xhCAAAAIBAJ&sjid=S6oMAAAAIBAJ&pg=4742,218227&dq=south+texas+project+protests&hl=en. Last accessed on October 1, 2012.

"Hundreds arrested in protest." 1979. *Beaver County Times.* http://news.google.com/newspapers?id=dIIkAAAAIBAJ&sjid=9NkFAAAAIBAJ&pg=1340,263988&dq=south-texas-nuclear&hl=en. Last accessed on September 28, 2012.

"Texas nuclear project ailing." 1979. *Anchorage Daily News*. http://news.google.com/newspapers?id=wzweAAAAIBAJ&sjid=VsAEAAAAIBAJ&pg=3644,1175858&dq=south-texas-nuclear&hl=en. Last accessed on September 28, 2012.

Federalism

Stepan, Alfred C. 1999. "Federalism and Democracy: Beyond the U.S. Model." *Journal of Democracy*10(4): 19–34.

Public Engagement

"Pro-nuclear Forces Win in Austin." 1979. *Anchorage Daily News*. http://news.google.com/newspapers?id=wzweAAAAIBAJ&sjid=VsAEAAAAIBAJ&pg=3944,1088999&dq=south+texas+project&hl=en. Last accessed on October 1, 2012.

"Texas nuclear project ailing." 1979. *Anchorage Daily News*. http://news.google.com/newspapers?id=wzweAAAAIBAJ&sjid=VsAEAAAAIBAJ&pg=3644,11 75858&dq=south-texas-nuclear&hl=en. Last accessed on September 28, 2012.

St. Lucie Plant

Federalism

Stepan, Alfred C. 1999. "Federalism and Democracy: Beyond the U.S. Model." *Journal of Democracy* 10(4): 19–34.

Sundesert Nuclear Power Plant

Activism

Angle, Martha, and Robert Walters. 1979. "Harassment Means Used on Anti-nuke Protestors." *The Tuscaloosa News*. http://news.google.com/newspapers?id=ly8dAAAAIBAJ&sjid=Bp4EAAAAIBAJ&pg=4544,1071851&dq=sundesert+nuclear+power+plant+protests&hl=en. Last accessed on October 1, 2012.

"Panel action kills planned Sundesert nuclear plant." 1978. *The Modesto Bee*. http://news.google.com/newspapers?id=fkohAAAAIBAJ&sjid=LIEFAAAAIBAJ&pg=4007,5323632&dq=sundesert-nuclear-plant&hl=en. Last accessed on October 1, 2012.

Stall, Bill. 1978. "Governor Opposed: Plant Symbol of Nuke Battle." *The Spokesman-Review*. http://news.google.com/newspapers?id=e0pOAAAAIBAJ&sjid=Bu4DAAAAIBAJ&pg=6715,2816553&dq=sundesert-nuclear-power-plant+public+opposition&hl=en. Last accessed on September 28, 2012.

Federalism

Stepan, Alfred C. 1999. "Federalism and Democracy: Beyond the U.S. Model." *Journal of Democracy* 10(4): 19–34.

Public Engagement

Stall, Bill. 1978. "Governor Opposed Plant Symbol of Nuke Battle." *The Spokesman-Review.* http://news.google.com/newspapers?id=e0pOAAAAIBAJ&sjid=Bu4DAAAAIBAJ&pg=6715,2816553&dq=sundesert+nuclear+power+plant+opposition&hl=en. Last accessed on October 1, 2012.

"Panel action kills planned Sundesert nuclear plant." 1978. *The Modesto Bee.* http://news.google.com/newspapers?id=fkohAAAAIBAJ&sjid=LIEFAAAAIBAJ&pg=4007,532 3632&dq=sundesert-nuclear-plant&hl=en. Last accessed on October 1, 2012.

Appendix B: Survey Data Sources and Question Wording

The analysis in Section 3 was based on a series of public opinion surveys taken over many years by many different organizations. Using the iPoll database, we gathered questions related to nuclear energy that spanned the years 1973 to 2011. These questions were placed into categories (such as more plants) based on similarity in question wording and these categories formed the basis of the nuclear risk and nuclear acceptability scales. This Appendix provides details on the categories including specific question wording, organization that administered the survey, and the years in which the questions were asked. The first part covers the nuclear risk scale and the second covers the acceptability scale.

The Nuclear Risk Scale

Nuclear Accident

Associated Press

(Here are some statements often made about nuclear power. Tell me whether you generally agree or disagree with each.)...Nuclear power is unsafe with great danger of accidents.

Years: 1989, 1999, 2001

CBS

How concerned are you that a major accident might occur at a nuclear power plant in the United States-- would you say you are very concerned, somewhat concerned, or not very concerned, or not at all concerned?

Year: 2011

Nuclear Safety

Gallup

Do you feel that nuclear power plants operating today are safe enough with the present safety regulations, or do you feel that their operations should be cut back until more strict regulations can be put into practice?

Years: 1976, 1979, 1980, 2011

Harris

How safe are nuclear power plants? Are they...very safe, somewhat safe, or not so safe? Years: 1976, 1982, 1988

Nuclear Energy Institute

Thinking about the nuclear energy plants that are operating now, how safe do you regard these plants? Please think of a scale from 1 to 7, where 1 means 'very unsafe' and 7 means 'very safe'. The safer you think they are, the higher the number you would give.

Years: 2002, 2003, 2004, 2005, 2006, 2007, 2008, 2009, 2010, 2011

Nuclear Risk

Roper

(On another subject, we hear a lot these days about things that can be risks to people's health or safety. Here is a list of a number of them. Would you read over that list, and then for each one tell me whether it is something you think involves a high degree of risk to a person, or involves a moderate risk, or involves only a minor risk?)...Living near a nuclear power plant

Years: 1979, 1981, 1985, 1986

Nimby Safe

Roper

Actually, there are differences in opinion about how safe atomic energy plants are. Some people say they are completely safe, while others say they present dangers and hazards. How do you feel that it would be safe to have an atomic energy plant someplace near here, or that it would present dangers?

Years: 1973, 1974, 1975, 1976, 1977, 1979, 1980, 1982, 1983

Nimby Danger

Trendex

Could you please tell me how much danger you feel there is of an accident or mishap in the following situations....Do you think there is great danger, some danger, or very little danger in...living near an atomic energy plant?

Years: 1972, 1976, 1977, 1978, 1979, 1982

The Nuclear Acceptability Scale

More Plants

Cambridge

Do you generally favor or oppose the construction of more nuclear power plants?

Years: 1976, 1977

(I'm going to read you a list of proposals for dealing with the energy crisis, and I'd like you to tell me whether you generally favor or oppose each one.)...Building more nuclear power plants

Years: 1977, 1978, 1979, 1980, 1981, 1982, 1983, 1984, 1985, 1986, 1987, 1988, 1989, 1990

How do you feel about building more nuclear power plants? In general, do you strongly favor, somewhat favor, somewhat oppose, or strongly oppose building more nuclear power plants?

Year: 1981

Gallup

In order to meet the future power needs of the nation, how important do you feel it is to have more nuclear power plants? extremely important, somewhat important, not too important, or not at all important?

Years: 1976, 1979, 1986, 1999

There various proposals being discussed in this country today. Would you tell me whether you generally favor or generally oppose each of these proposals? More nuclear power plants

Years: 1980, 1990

Harris

For each, tell me if you would favor or oppose taking that step to help solve the energy shortage?...Speed up the development of more nuclear power plants from 10 years to 6 years

Years: 1973, 1974, 1976, 1977

Do you favor or oppose the building of more nuclear power plants in the United States? Years: 1976, 1978, 1979, 1980, 1983, 1988

Nuclear Electricity

Associated Press

Do you support the use of nuclear power to generate electricity or not, or don't you have an opinion either way?

Years: 1986, 1989, 1999, 2001

Gallup

Do you strongly favor, somewhat favor, somewhat oppose, or strongly oppose nuclear energy as one of the ways to provide electricity for the United States?

Years: 1994, 2001

Roper

Generally speaking, do you approve or disapprove of using nuclear energy to produce electric power? Years: 1979, 1986

Nuclear Increase

Roper

Go into a greatly increased program to develop atomic energy. Is that something you think we should do or not do?

Years: 1973, 1977, 1979, 1980, 1981

ORC

As part of the solution to our energy problems, do you favor or oppose the increased use of nuclear power to generate electricity?

Years: 1979, 1980, 1983, 1984

Government Promotion

Pew

As I read some possible government policies to address America's energy supply, tell me whether you would favor or oppose each. First, would you favor or oppose the government...Promoting the increased use of nuclear power

Years: 2005, 2006, 2008, 2009, 2010, 2011

APPENDIX C: ENERGY AND ENVIRONMENT SURVEY DATA SUMMARY

Survey implemented via Internet, 1–5 June 2012; Average time for completion 40 minutes.

Sample size is 2017 completed online interviews.

e1_age How old are you?

	Mean
2012 web	45.9

e2_edu What is the highest level of education you have completed?

%	**2012 web**
1. < High school graduate	1
2. High school graduate	19
3. Some college/vocational sch	33
4. College graduate	29
5. Some graduate work	5
6. Master's degree	9
7. Doctorate (of any type)	3
8. Other degree	<1

e3_gend Are you male or female?

	Female	Male
%	0	1
12 web	51.0	49.0

e4_state Using the dropdown list, please select the state where your primary residence is located.

%	Northeast	Midwest	South	West
12 web	19	23	34	24

e5_zip What is the five digit zip code at your residence? (This information will only be used to compare grouped differences, not to identify you.) [verbatim]

e6_rural Which of the of the following categories best describes the location of your primary residence?

1 Urban: within the incorporated boundaries of a city or town that provides emergency services such as fire, rescue, and storm warnings for your residence
2 Suburban: near or in a suburb or town that provides emergency services such as fire, rescue, and storm warnings for your residence
3 Rural: not within the incorporated boundaries of a city or town; emergency services such as fire, rescue, and storm warnings for your residence usually are provided by county, state, or federal entities

	Urban	Suburban	Rural
%	**1**	**2**	**3**
12 web	40	46	15

e7_now Please indicate which of the following statements applies to you.

0 – I am completing this survey from my primary residence.

1 – I am completing this survey from a location that is not my primary residence.

	Primary Residence	Not Primary Residence
%	**0**	**1**
12 web	86	14

Now I want to ask you some questions about important issues facing policy makers in the U.S. today.

For each of the following issues, please rate your level of concern about the issue using a scale from zero to ten, where zero means you are not at all concerned and ten means you are extremely concerned. How concerned are you about: [e8–e12 Randomized]

e8_worry1 Threats to national security, including terrorism?

Not at All Concerned									Extremely Concerned			
%	**0**	**1**	**2**	**3**	**4**	**5**	**6**	**7**	**8**	**9**	**10**	**Mean**
12 web	2	2	2	4	5	9	9	13	16	13	26	7.35

e9_worry2 The delivery and cost of healthcare in the U.S.?

Not at All Concerned									Extremely Concerned			
%	**0**	**1**	**2**	**3**	**4**	**5**	**6**	**7**	**8**	**9**	**10**	**Mean**
12 web	1	0	1	1	1	4	5	10	15	18	43	8.48

e10_worry3 The availability and cost of energy in the U.S.?

Not at All Concerned									Extremely Concerned			
%	**0**	**1**	**2**	**3**	**4**	**5**	**6**	**7**	**8**	**9**	**10**	**Mean**
12 web	1	0	1	1	2	5	7	14	19	18	33	8.12

e11_worry4 The effects of human activities on the environment? (NOTE: wording change in 09)

Not at All Concerned									Extremely Concerned			
%	**0**	**1**	**2**	**3**	**4**	**5**	**6**	**7**	**8**	**9**	**10**	**Mean**
12 web	3	2	2	2	3	9	10	15	16	14	24	7.30

e12_worry5 The state of the economy, including jobs and inflation?

Not at All Concerned									Extremely Concerned			
%	**0**	**1**	**2**	**3**	**4**	**5**	**6**	**7**	**8**	**9**	**10**	**Mean**
12 web	1	0	0	1	1	3	4	9	14	20	48	8.79

The next several questions ask about your views on energy and environmental issues. These questions concern your perceptions and beliefs, so don't worry about being right or wrong when providing your answers.

e13_futr Using a scale from zero to ten, where zero means you are *not at all confident* and ten means you are *completely confident*, how confident are you that there will be adequate sources of energy to meet the energy needs of the U.S. during the next 20 years? Please think about U.S. energy needs overall, including transportation, heating, electricity, and other energy requirements when considering your answer.

Not at All Confident									Completely Confident			
%	**0**	**1**	**2**	**3**	**4**	**5**	**6**	**7**	**8**	**9**	**10**	**Mean**
12 web	3	2	4	7	9	18	13	16	13	5	10	5.96

e14_egpol As you may know, U.S. energy policies generally deal with such issues as the sources and adequacy of energy supplies, the costs of various types of energy, and the environmental implications of using energy. Using a scale from zero to ten, where zero means *not at all satisfied* and ten means *completely satisfied*, how satisfied are you with current U.S. energy policies overall?

Not at All Satisfied									Completely Satisfied			
%	**0**	**1**	**2**	**3**	**4**	**5**	**6**	**7**	**8**	**9**	**10**	**Mean**
12 web	9	4	8	14	13	23	12	9	5	2	2	4.37

e15_nature On a scale from zero to ten, where zero means that nature is *robust and not easily damaged* and ten means nature is *fragile and easily damaged*, how do you view nature?

Robust and *Not* Easily Damaged									**Fragile and *Is* Easily Damaged**			
%	**0**	**1**	**2**	**3**	**4**	**5**	**6**	**7**	**8**	**9**	**10**	**Mean**
12 web	3	1	4	6	7	17	11	17	15	7	13	6.38

As you may know, the issue of global climate change has been the subject of public discussion over the last few years.

e16_gcc In your view, are greenhouse gases, such as those resulting from the combustion of coal, oil, natural gas, and other materials causing average global temperatures to rise?

	Are Not	**Are**
%	**0**	**1**
12 web	28	72

e17_gcccert On a scale from zero to ten, where zero means *not at all certain* and ten means *completely certain*, how certain are you that greenhouse gases <are/are not> (from e16) causing average global temperatures to rise?

Not at All Certain									**Completely Certain**			
%	**0**	**1**	**2**	**3**	**4**	**5**	**6**	**7**	**8**	**9**	**10**	**Mean**
12 web	2	1	1	4	4	14	15	19	17	9	13	6.76

e18_gccrsk On the scale from zero to ten, where zero means *no risk* and ten means *extreme risk*, how much risk do you think global warming poses for people and the environment?

No Risk										**Extreme Risk**		
%	**0**	**1**	**2**	**3**	**4**	**5**	**6**	**7**	**8**	**9**	**10**	**Mean**
12 web	3	2	3	5	5	12	11	16	15	10	17	6.70

The next set of questions concerns all kinds and uses of energy, including electricity for homes and businesses; gas, oil, and coal for heating; and transportation fuels, such as gasoline and diesel.

Considering the effects of both normal operations and potential accidents, how do you rate the risks to society and the environment from each of the following sources of energy using a scale from zero to ten, where zero means *no risk* and ten means *extreme risk*? [e19–e21 Randomized]

e19_ersk1 The risks from fossil fuels, such as coal, oil, and natural gas?

No Risk									Extreme Risk		
%	**0**	**1**	**2**	**3**	**4**	**5**	**6**	**7**	**8**	**9**	**10**	**Mean**
12 web	3	3	5	6	7	14	14	13	16	9	10	6.12

e20_ersk2 The risks from nuclear power plants?

No Risk									Extreme Risk		
%	**0**	**1**	**2**	**3**	**4**	**5**	**6**	**7**	**8**	**9**	**10**	**Mean**
12 web	1	3	4	5	5	12	10	12	14	12	21	6.88

e21_ersk3 The risks from renewable sources of energy, such as from hydroelectric dams, solar power, and wind generation?

No Risk									Extreme Risk		
%	**0**	**1**	**2**	**3**	**4**	**5**	**6**	**7**	**8**	**9**	**10**	**Mean**
12 web	18	19	16	10	6	10	7	4	4	3	3	3.09

Please respond to the following statements using a continuous scale from one to seven, where one means *strongly disagree* and seven means *strongly agree*.

***e22_nucgg*:** Nuclear power plants produce significant amounts of greenhouse gases.

Strongly Disagree			Strongly Agree		
%	**1**	**2**	**3**	**4**	**5**	**6**	**7**	**Mean**
12 web	13	10	13	28	18	9	10	3.95

e23_depd Using a scale from zero to ten, where zero means *not at all important* and ten means *extremely important*, how important is it to reduce U.S. dependence on foreign sources of energy of all types?

Not at All Important										**Extremely Important**		
%	**0**	**1**	**2**	**3**	**4**	**5**	**6**	**7**	**8**	**9**	**10**	**Mean**
12 web	1	0	1	1	2	6	5	11	18	13	42	8.33

Split Design Group-A (50%): Total Energy

Now think about the overall mix of all sources of energy for the U.S. We currently get about 83 percent of our energy from fossil fuels, 9 percent from nuclear energy, 3 percent from hydroelectric dams on rivers and lakes, and 5 percent from other renewable sources (wood, biofuels, wind, waste products, geothermal, and solar).We want to know approximately what percentage of the total U.S. energy supply over the next 20 years you would like to see come from each of these four sources. [Randomized]

e24A_foss What percent of our energy should come from fossil fuels, which currently provide about 83% of total U.S. energy? [verbatim]

% Fossil Fuels (Mean)	
12 web-A	32

e25A_nuc What percent of our energy should come from nuclear energy, which currently provides about 9% of total U.S. energy? [verbatim]

% Nuclear (Mean)	
12 web-A	13

e26A_hydro What percent of our energy should come from hydroelectric dams on rivers and lakes, which currently provide about 3% of total U.S. energy? [verbatim]

% Hydroelectric (Mean)	
12 web-A	20

e26a_A_renew What percent of our energy should come from renewable sources (wood, biofuels, wind, waste products, geothermal, and solar), which currently provide about 5% of total U.S. energy? [verbatim]

% Other Renewables (Mean)	
12 web-A	35

Split Design: Group-B (50%): Total Electricity

Now think about the overall mix of all sources of electricity for the U.S. We currently get about 69 percent of our electricity from fossil fuels, 20 percent from nuclear energy, 6 percent from hydroelectric dams on rivers and lakes, and 6 percent from other renewable sources (wood, wind, waste products, geothermal, solar, and other). We want to know approximately what percentage of the total U.S. electricity supply over the next 20 years you would like to see come from each of these four sources.

e24B_foss What percent of our electricity should come from fossil fuels, which currently provide about 69% of total U.S. electricity? [verbatim]

% Fossil Fuels (Mean)	
12 web-B	25

e25B_nuc What percent of our electricity should come from nuclear energy, which currently provides about 20% of total U.S. electricity? [verbatim]

% Nuclear (Mean)	
12 web-B	18

e26B_hydro What percent of our electricity should come from hydroelectric dams on rivers and lakes, which currently provide about 6% of total U.S. electricity? [verbatim]

% Hydroelectric (Mean)	
12 web-B	22

e26a_B_renew What percent of our electricity should come from renewable sources (wood, biofuels, wind, waste products, geothermal, and solar), which currently provide about 6% of total U.S. electricity? [verbatim]

% Other Renewables (Mean)	
12 web-B	36

End Split A/B

[Arguments in Random Order]

Some people argue that regardless of the future mix of energy sources, we must also significantly reduce energy consumption.

Some people think that significantly reducing energy consumption limits economic growth and is not practical.

e27_needs Considering both arguments and using the zero-to-ten scale below, where zero means *place all efforts on reducing energy consumption and ten means place all efforts on developing the energy mix you previously identified*, what strategy would you prefer? Notice that when you select a response, the resulting balance is shown in the two boxes.

All Efforts on Conservation										**All Efforts on Development**		
%	**0**	**1**	**2**	**3**	**4**	**5**	**6**	**7**	**8**	**9**	**10**	**Mean**
12 web	1	1	2	4	7	28	14	18	13	5	8	6.18

There is another important debate about the energy future that we want you to consider. [Randomized]

Some people *oppose* further developing U.S. deposits of oil and gas. They argue that doing so increases greenhouse gas emissions, harms the environment, and reduces the economic incentives for developing alternative sources of energy that are cleaner.

Some people *support* further developing U.S. deposits of oil and gas. They argue that doing so keeps energy prices lower, reduces dependence on foreign sources, and gains time for developing alternative sources of energy that are cleaner.

e28_explore Considering both arguments and using a scale from one to seven where one means strongly oppose and seven means *strongly support*, how do you feel about further exploring and developing U.S. deposits of oil and gas?

Strongly Oppose						Strongly Support		
%	1	2	3	4	5	6	7	Mean
12 web	5	6	11	23	23	12	21	4.72

New drilling and extraction techniques are making it possible to recover large U.S. reserves of oil and natural gas that cannot be extracted using conventional drilling methods. One of the new techniques involves drilling into underground shale deposits (a type of sedimentary rock) and injecting pressurized liquids that cause the shale to fracture, releasing oil and gas. This technique is sometimes called "fracking."

[Arguments in Random Order]

Opponents of fracking argue that the liquids injected into the shale deposits may contaminate sources of water and that the fracturing of the shale creates geologic instability. They also argue that extracting more U.S. oil and gas keeps us dependent on fossil fuels whose use contributes to global warming and delays our transition to alternative sources of energy.

Supporters of fracking argue that it can be done without polluting underground water or creating instability and that we need to use U.S. oil and gas resources to help control energy costs and reduce dependence on foreign energy. They argue that producing more U.S. oil and gas will help meet domestic energy needs while we are developing alternative sources.

e29_frack on a scale from one to seven, where one means *strongly oppose* and seven means *strongly support*, how do you feel about using fracking methods to extract U.S. oil and natural gas resources?

Strongly Oppose						Strongly Support		
%	1	2	3	4	5	6	7	Mean
12 web	10	10	15	25	20	10	11	4.09

Canada has large deposits of oil sands from which oil can be extracted. A Canadian company wants to build an underground pipeline network to carry that oil to refineries in Illinois and Texas. The project is known as the Keystone XL Pipeline. The portion of the pipeline that would be in Canada has been approved by the Canadian government. The portions of the pipeline that would be in the U.S. have not yet been approved by the U.S. government.

[Arguments in Random Order]

Opponents of the Keystone XL Pipeline argue that such a lengthy pipeline poses unacceptable environmental risks, including the potential for leaks that may pollute underground aquifers that provide crucial sources of water. They also argue that this kind of oil extracted from sands is especially "dirty" and contributes to global warming.

Supporters of the Keystone XL Pipeline argue that it can be built and operated without causing unacceptable environmental risks, and that the pipeline will create thousands of American jobs and help reduce U.S. dependence on oil from the Middle East.

e30_pipe: Again using a scale from one to seven, where one means strongly oppose and seven means strongly support, how do you feel about building the Keystone XL Pipeline in the U.S.?

Strongly Oppose						**Strongly Support**		
%	**1**	**2**	**3**	**4**	**5**	**6**	**7**	**Mean**
12 web	7	6	11	23	20	13	20	4.59

The next set of questions focuses specifically on the possible risks and benefits of nuclear energy.

First we want to know about your beliefs concerning some of the possible risks associated with nuclear energy use in the U.S. Please consider both the likelihood of a nuclear event occurring and its potential consequences when evaluating the risk posed by each of the following on a scale from zero to ten where zero means *no risk* and ten means *extreme risk.* [e31–e34 Randomized]

e31_nrsk1 An event at a U.S. nuclear power plant within the next 20 years that results in the release of large amounts of radioactivity.

No Risk										**Extreme Risk**		
%	**0**	**1**	**2**	**3**	**4**	**5**	**6**	**7**	**8**	**9**	**10**	**Mean**
12 web	2	5	6	7	6	11	10	12	12	10	19	6.42

e32_nrsk2 An event during the transportation or storage of used nuclear fuel from nuclear power plants in the U.S. within the next 20 years that results in the release of large amounts of radioactivity.

No Risk									Extreme Risk		
%	**0**	**1**	**2**	**3**	**4**	**5**	**6**	**7**	**8**	**9**	**10**	**Mean**
12 web	2	6	6	6	6	15	9	13	13	8	17	6.19

e33_nrsk3 A terrorist attack at a U.S. nuclear power plant within the next 20 years that results in the release of large amounts of radioactivity.

No Risk										Extreme Risk		
%	**0**	**1**	**2**	**3**	**4**	**5**	**6**	**7**	**8**	**9**	**10**	**Mean**
12 web	2	4	6	5	5	13	9	13	12	10	21	6.57

e34_nrsk4 The diversion of nuclear fuel from a nuclear power plant in the U.S. within the next 20 years for the purpose of building a nuclear weapon.

No Risk										Extreme Risk		
%	**0**	**1**	**2**	**3**	**4**	**5**	**6**	**7**	**8**	**9**	**10**	**Mean**
12 web	4	8	8	7	7	14	9	12	11	6	15	5.70

Now we want to know about your beliefs concerning some of the possible benefits associated with nuclear energy use in the U.S. Please evaluate the benefits associated with each of the following on a scale from zero to ten, where zero means *not at all beneficial* and ten means *extremely beneficial.* [e35–e38 Randomized]

e35_nben1 Fewer overall greenhouse gas emissions because nuclear energy production does not create greenhouse gases.

Not At All Beneficial										Extremely Beneficial		
%	**0**	**1**	**2**	**3**	**4**	**5**	**6**	**7**	**8**	**9**	**10**	**Mean**
12 web	3	1	2	3	6	16	11	15	16	11	15	6.74

e36_nben2 Reliable power because nuclear energy generates large amounts of electricity and is not affected by weather conditions, such as low rainfall or no wind.

Not At All Beneficial										**Extremely Beneficial**		
%	**0**	**1**	**2**	**3**	**4**	**5**	**6**	**7**	**8**	**9**	**10**	**Mean**
12 web	2	1	2	3	5	16	11	15	17	12	17	6.89

e37_nben3 Greater U.S. energy independence because nuclear energy production does not require oil or gas from foreign sources.

Not At All Beneficial										**Extremely Beneficial**		
%	**0**	**1**	**2**	**3**	**4**	**5**	**6**	**7**	**8**	**9**	**10**	**Mean**
12 web	2	1	2	3	5	13	10	15	16	13	20	7.04

e38_nben4 Reduced environmental damage because of less need for mining coal or extracting oil and gas.

Not At All Beneficial										**Extremely Beneficial**		
%	**0**	**1**	**2**	**3**	**4**	**5**	**6**	**7**	**8**	**9**	**10**	**Mean**
12 web	2	1	3	4	5	15	11	16	15	12	15	6.75

Now please consider the overall balance of these possible risks and benefits of nuclear energy in the U.S.

e39_riskben Using a scale from one to seven, where one means the risks of nuclear energy far outweigh its benefits, four means the risks and benefits are equally balanced, and seven means the benefits of nuclear energy far outweigh its risks, how do you rate the overall balance of the risks and benefits of nuclear energy in the U.S.? Remember, you can choose any number from one to seven.

Risks > Benefits			**Risks/Benefits Balanced**			**Benefits > Risks**		
%	**1**	**2**	**3**	**4**	**5**	**6**	**7**	**Mean**
12 web	6	8	15	35	19	11	7	4.11

e40_new1 Using a scale from one to seven, where one means *strongly oppose* and seven means *strongly support*, how do you feel about constructing additional nuclear reactors at the sites of existing nuclear power plants in the U.S.?

Strongly Oppose						**Strongly Support**		
%	**1**	**2**	**3**	**4**	**5**	**6**	**7**	**Mean**
12 web	11	9	15	27	19	10	9	3.98

e41_new2 Using the same scale from one to seven, where one means *strongly oppose* and seven means *strongly support*, how do you feel about constructing additional nuclear power plants at new locations in the U.S.?

Strongly Oppose						**Strongly Support**		
%	**1**	**2**	**3**	**4**	**5**	**6**	**7**	**Mean**
12 web	15	11	15	21	17	10	10	3.84

e42_near To the best of your knowledge, is your primary residence located within approximately 100 miles of an operating nuclear power plant?

	No	**Yes**	**Don't Know**		
%	**0**	**1**	**2**	**Correct**	**Incorrect/DK**
12 web	43	33	25	45	55

e43_disp As nuclear fuel is used to generate electricity, it becomes contaminated with radioactive byproducts. When it can no longer efficiently produce electricity, it is called "used" or "spent" nuclear fuel. To the best of your knowledge, what is currently being done with most of the used nuclear fuel produced in the U.S.? [Randomized]

%	**2012 web**
1. Stored in cooling pools or special containers at nuclear power plants throughout the US	39
2. Shipped to Nevada and stored in a facility deep underground	22
3. Chemically reprocessed and reused	15
4. Shipped to regional storage sites	24

e44_casks: To the best of your knowledge, is used nuclear fuel being stored above ground at any nuclear power plant within your state?

	No	Yes	Don't Know		
%	0	1	2	Correct	Incorrect/DK
12 web	25	12	63	14	86

Split Design: Group-C (33%): Same As 2011

Used nuclear fuel is highly radioactive and must be safeguarded for thousands of years or chemically reprocessed. If it is reprocessed, the uranium can be separated from the waste and reused to make new fuel rods for generating electricity, but the remaining elements are highly radioactive for a very long time and must be safeguarded and isolated from the environment for thousands of years.

In 2010 the government halted construction of a deep underground facility inside Yucca Mountain in Nevada that had been intended for long-term disposition of used nuclear fuel, and very little used nuclear fuel is being reprocessed in the U.S.

Currently, U.S. used nuclear fuel is being temporarily stored at over 100 sites in 39 states. Most of it is stored at nuclear power plants where it is placed in specialized concrete cooling pools. In some cases, the used fuel is transferred to specialized concrete casks stored above ground near the nuclear power plant. At each site, the cooling pools and storage casks are protected at all times by security forces. Some people think this is an acceptable solution for the foreseeable future, while others think such practices are risky and other options need to be adopted.

[Arguments in Random Order]

Opponents argue that some nuclear power plants where used nuclear fuel is stored are near rivers, oceans, and large population centers. On rare occasions used fuel has leaked radiation into the cooling pools. Moreover, the cooling pools and containers are located at ground level, and therefore might be vulnerable to terrorists. They note that these storage practices do not provide a permanent solution for managing used nuclear fuel.

Supporters argue that transporting used nuclear fuel by train or truck to consolidated storage facilities is risky, that storing used nuclear fuel at nuclear power plants is less expensive than consolidated storage, and that it buys time for finding future solutions. Moreover, storage at nuclear power plants has not caused any accidents in the United States that have exposed the public to radiation.

Split Design: Group-D (33%): Adds BRC Recommendations

Used nuclear fuel is highly radioactive and must be safeguarded for thousands of years or chemically reprocessed. If it is reprocessed, the uranium can be separated from the waste and reused to make new fuel rods for generating electricity, but the remaining elements are highly radioactive for a very long time and must be safeguarded and isolated from the environment for thousands of years.

In 2010 the government halted construction of a deep underground facility inside Yucca Mountain in Nevada that had been intended for long-term disposition of used nuclear fuel, and very little used nuclear fuel is being reprocessed in the U.S.

After two years of study, in January 2012 the President's "Blue Ribbon Commission on America's Nuclear Future" recommended quick efforts to build one or more underground nuclear repositories for permanent storage and disposal of used nuclear fuel. Another recommendation was to build one or more interim sites for managing and temporarily storing used nuclear fuel. The Commission also recommended making preparations for transporting nuclear materials to those storage and disposal facilities.

Currently, U.S. used nuclear fuel is being temporarily stored at over 100 sites in 39 states. Most of it is stored at nuclear power plants where it is placed in specialized concrete cooling pools. In some cases, the used fuel is transferred to specialized concrete casks stored above ground near the nuclear power plant. At each site, the cooling pools and storage casks are protected at all times by security forces. Some people think this is an acceptable solution for the foreseeable future, while others think such practices are risky and other options need to be adopted.

[Arguments in Random Order]

Opponents argue that some nuclear power plants where used nuclear fuel is stored are near rivers, oceans, and large population centers. On rare occasions used fuel has leaked radiation into the cooling pools. Moreover, the cooling pools and containers are located at ground level, and therefore might be vulnerable to terrorists. They argue that these storage practices do not provide a permanent solution for managing used nuclear fuel.

Supporters argue that transporting used nuclear fuel by train or truck to consolidated storage facilities is risky, that storing used nuclear fuel at nuclear power plants is less expensive than consolidated storage, and that it buys time for finding future solutions. Moreover, storage at nuclear power plants has not caused any accidents in the United States that have exposed the public to radiation.

Split Design: Group-E (33%): Adds Flooding and Stranded Fuel; New Baseline

Used nuclear fuel is highly radioactive and must be safeguarded for thousands of years or chemically reprocessed. If it is reprocessed, the uranium can be separated from the waste and reused to make new fuel rods for generating electricity, but the remaining elements are highly radioactive for a very long time and must be safeguarded and isolated from the environment for thousands of years.

In 2010 the government halted construction of a deep underground facility inside Yucca Mountain in Nevada that had been intended for permanent storage and disposal of used nuclear fuel, and very little used nuclear fuel is being reprocessed in the U.S.

After two years of study, in January 2012 the President's "Blue Ribbon Commission on America's Nuclear Future" recommended quick efforts to build one or more underground nuclear repositories for permanent storage and disposal of used nuclear fuel. Another recommendation was to build one or more interim sites for managing and temporarily storing used nuclear fuel. The Commission also recommended making preparations for transporting nuclear materials to those storage and disposal facilities.

Currently, U.S. used nuclear fuel is being temporarily stored at over 100 sites in 39 states. Most of it is stored at nuclear power plants where it is placed in specialized concrete cooling pools. In some cases, the used fuel is transferred to specialized concrete casks stored above ground near the nuclear power plant. At each site, the cooling pools and storage casks are protected at all times by security forces. This poses a problem at nine sites where nuclear power plants have been shut down but the used nuclear fuel stored there continues to require expensive security measures that otherwise would not be needed. Some people think that temporarily storing used nuclear fuel at existing sites is an acceptable solution for the foreseeable future, while others think such practices are risky and other options need to be adopted.

[Arguments in Random Order]

Opponents argue that some nuclear power plants where used nuclear fuel is stored are near rivers and oceans where flooding is possible. And some of these nuclear power plants are near large population centers. On rare occasions used nuclear fuel has leaked radiation into the cooling pools. Moreover, the cooling pools and storage casks are located at ground level and therefore might be vulnerable to terrorists. Opponents argue that these storage practices do not provide a permanent solution for managing used nuclear fuel.

Supporters argue that transporting used nuclear fuel by train or truck to consolidated storage facilities is risky, that storing used nuclear fuel at nuclear power plants is less expensive than consolidated storage, and that it buys time for finding future solutions. Moreover, supporters argue that temporarily storing used nuclear fuel at nuclear power plants has not caused any accidents in the United States that have exposed the public to radiation.

e45_opt1 Using a scale from one to seven, where one means strongly oppose and seven means strongly support, how do you feel about the current practice of storing used nuclear fuel at or near nuclear power plants?

Strongly Oppose						**Strongly Support**		
%	**1**	**2**	**3**	**4**	**5**	**6**	**7**	**Mean**
12 web-C	12	13	20	30	18	5	2	3.56
12 web-D	13	12	19	31	15	6	3	3.53
12 web-E	13	13	23	29	15	5	3	3.44

[12 web-C vs. D: $p = .7585$][12 web-C vs. E: $p = .1511$] [12 web-D vs. E: $p = .2759$]

End Split C/D/E

Now consider the number of storage sites for used nuclear fuel. While nuclear power plants will continue to store some used nuclear fuel in their cooling pools, much of the radioactive materials currently at more than 100 temporary storage sites in 39 states might be consolidated at a smaller number of interim storage facilities. Once it is consolidated, the used nuclear fuel can more easily be secured and protected from attack. The fewer the number of interim storage facilities for used nuclear fuel, the less complex are the political and legal obstacles for finding communities willing and able to host the facilities. At the same time, a larger number of interim storage facilities for used nuclear fuel would reduce the distances radioactive materials must be transported by train or truck, and would also reduce the number of communities through which the transport routes would pass.

Please respond to the three following policy options on a scale from one to seven, where one means *strongly oppose* and seven means *strongly support.* [e46–e48 Randomized]

e46_nmbrs1 After used nuclear fuel is removed from cooling pools, continue the current practice of temporarily storing it near ground level at designated nuclear power plants in 39 states. This option does not require additional transportation of radioactive materials by train or truck, but it is not without political and legal obstacles. Some states are suing the federal government to end temporary storage practices at nuclear power plants.

Strongly Oppose						**Strongly Support**		
%	**1**	**2**	**3**	**4**	**5**	**6**	**7**	**Mean**
12 web	8	11	16	32	19	9	5	3.92

e47_nmbrs2 Construct several interim storage facilities that would be easier to secure and could store used nuclear fuel safely up to a hundred years, which is longer then envisioned for current storage at nuclear power plants. Eventually, the materials would need to be moved to a permanent nuclear repository. This option initially requires transporting used nuclear fuel by train or truck over moderate distances and is likely to generate political and legal opposition.

Strongly Oppose						**Strongly Support**		
%	**1**	**2**	**3**	**4**	**5**	**6**	**7**	**Mean**
12 web	6	9	13	29	25	11	7	4.17

e48_nmbrs3 Construct two large nuclear repositories (one in the western U.S. and one in the east) that can be most easily secured and would provide permanent storage and disposal of used nuclear fuel for thousands of years. This option requires transporting used nuclear fuel by train or truck over longer distances and is likely to generate political and legal opposition.

Strongly Oppose						**Strongly Support**		
%	**1**	**2**	**3**	**4**	**5**	**6**	**7**	**Mean**
12 web	6	8	11	24	20	17	14	4.50

[e46 vs. e47: $p < .0001$] [e46 vs. e48: $p < .0001$] [e47 vs. e48: $p < .0001$]

Next we want you to consider the issue of reprocessing, which involves the chemical separation of radioactive materials in used nuclear fuel. After reprocessing, most of the uranium and plutonium can be captured and reused to generate electricity, reducing the amount of uranium that must be mined in the U.S. or purchased from other countries. Remaining materials are radioactive and must be safeguarded and isolated from the environment.

Finally, reprocessing may also separate the plutonium which, like uranium, could be used to make nuclear weapons.

e49_reproc Using a scale from one to seven, where one means strongly oppose and seven means strongly support, how do you feel about the option for reprocessing used nuclear fuel?

Strongly Oppose						**Strongly Support**		
%	**1**	**2**	**3**	**4**	**5**	**6**	**7**	**Mean**
12 web	5	3	8	25	27	18	14	4.76

Split Design: Group-F (50%): No Mention of Nonproliferation

Now consider the issue of whether stored radioactive materials should be managed in a way that allows authorized personnel to gain access to them and retrieve the materials in the future, or that seeks to permanently block access to them.

One option is to build facilities where the stored materials are continuously monitored and can be retrieved for reprocessing, or possibly to make them less dangerous using future technological developments. This option requires greater security efforts and may be more vulnerable to attack or theft.

Another option is to attempt to seal off storage sites in such a way that people cannot readily gain access to the materials in the future. This option is more secure, but it also does not allow reprocessing or treatment by future technological advancements.

Split Design: Group-G (50%): Adds Nonproliferation

Now consider the issue of whether stored radioactive materials should be managed in a way that allows authorized personnel to gain access to them and retrieve the materials in the future, or that seeks to permanently block access to them.

One option is to build facilities where the stored materials are continuously monitored and can be retrieved for reprocessing, or possibly to make them less dangerous using future technological developments. This option requires greater security efforts and may be more vulnerable to attack or theft.

Another option is to attempt to seal off storage sites in such a way that people cannot readily gain access to the materials in the future. This option is more secure, and it makes it less likely that the materials could be used to make nuclear weapons in the future, but it also does not allow reprocessing or treatment by future technological advancements.

Using a scale from one to seven, where one means *strongly oppose* and seven means *strongly support*, please indicate how you feel about each of the following two options. [e50–e51 Randomized]

e50_retrieve1: Construct nuclear repositories so that stored materials are monitored and can readily be retrieved in the future.

Strongly Oppose						Strongly Support		
%	1	2	3	4	5	6	7	Mean
12 web-F	4	4	9	22	27	22	12	4.75
12 web-G	5	5	8	22	28	20	11	4.70

[12 web-F vs. G: $p = .3986$]

e51_retrieve2: Construct nuclear repositories so that stored materials are permanently sealed away and cannot readily be retrieved in the future.

Strongly Oppose						Strongly Support		
%	1	2	3	4	5	6	7	Mean
12 web-F	5	10	19	27	18	12	10	4.18
12 web-G	5	11	17	25	17	13	11	4.23

[12 web-F vs. G: $p = .4580$]

End Split F/G

Next we want you to consider the issue of storage depth. There are three general options. [Randomized]

One option is to store used nuclear fuel at or near the surface in concrete and steel structures. This allows monitoring and retrieval, but it is considered to provide a safe means to manage the materials for only about a hundred years.

One option is to build mine-like storage and disposal facilities that are deep underground. These can be constructed to allow materials to be retrieved, or they can be designed to permanently block access in the future. They are suitable for storage over thousands of years.

One option involves drilling multiple boreholes of about 1.5 feet in diameter and up to three miles deep. Used nuclear fuel would be stored in the deepest parts of the boreholes that are in bedrock. There is almost no chance that the materials could migrate into the surface environment over thousands of years, and they would be extremely difficult to retrieve after the boreholes are sealed.

Please respond to the three following policy options on a scale from one to seven, where one means *strongly oppose* and seven means *strongly support*. [e52–e54 Randomized]

e52_facility1 Construct storage facilities at or near the surface of the earth that are less permanent but allow retrieval for reprocessing, research, or other treatments.

Strongly Oppose						Strongly Support		
%	1	2	3	4	5	6	7	Mean
12 web	9	12	14	28	19	11	6	3.91

e53_facility2: Construct storage facilities underground that are like mines that could be either permanently sealed or could allow materials to be retrieved.

Strongly Oppose						Strongly Support		
%	1	2	3	4	5	6	7	Mean
12 web	5	5	7	24	25	21	14	4.78

e54_facility3: Construct very deep boreholes that afford permanent and safe disposal, but would make materials extremely difficult to be retrieved after the boreholes are sealed.

Strongly Oppose						Strongly Support		
%	1	2	3	4	5	6	7	Mean
12 web	8	12	16	26	18	13	8	4.03

Split Design: Group-H (75%): Received Estimated Distance to Nearest Unf Storage Site

Based on the location information you provided, we estimate that your primary residence is approximately [insert estimate] miles (straight line) from the nearest nuclear energy facility where used nuclear fuel is currently in temporary storage. (Our estimate could be wrong, but you can check by looking at this map showing where used nuclear fuel currently is being stored in the U.S.)

e54a_map By voluntarily selecting button, respondents see map of UNF storage sites.

	Did NOT View Map	DID View Map
%	**0**	**1**
12 web	39	61

Split Design: Group-I (25%): Received No Estimate of Distance or Opportunity to View Map (Skip to Next Question)

End Split H/I

Split Design: Group-J (50%): Long-Term Repository

For the next few questions, assume that construction of two underground mine-like repositories is being considered for long-term storage and disposal of used nuclear fuel. One would be in the eastern U.S., and the other in the west. Each of these sites would include secure surface buildings and a mine deep underground where radioactive materials could be isolated from people and the environment and could be designed to allow retrieval or to permanently seal away the materials. The facilities and the mines would be designed to meet all technical and safety requirements set by the U.S. Nuclear Regulatory Commission, the U.S. Environmental Protection Agency, and state regulatory agencies.

Split Design: Group-K (50%): Interim Storage Facilities

For the next few questions, assume that construction of one or more interim above-ground storage facilities is being considered where used nuclear fuel could be stored safely for up to a hundred years. Each of these sites would include secure surface facilities where used nuclear fuel could be consolidated and stored, and where the radioactive materials could cool and be prepared and packaged for later shipment to a permanent repository. These interim storage facilities would be designed to meet all technical and safety requirements set by the U.S. Nuclear Regulatory Commission, the U.S. Environmental Protection Agency, and state regulatory agencies.

e55J_mines / e55K_interim Using a scale from one to seven where one means strongly oppose and seven means strongly support, how do you feel about this option?

Strongly Oppose						Strongly Support		
%	1	2	3	4	5	6	7	Mean
12 web-J	4	3	10	27	31	15	11	4.65
12 web-K	6	5	12	33	26	10	7	4.29

[12 web-J vs. K: $p < .0001$]

Now we want you to consider how your support would be affected by more specific information. Please respond to each of the following questions on a scale from one to seven, where one means *strongly oppose* and seven means *strongly support*. [e56–e58 Randomized]

e56J_lab What would your level of support be if you learned that each of the permanent repositories for used nuclear fuel also would include a national research laboratory for studying ways to more safely and efficiently manage and dispose of nuclear materials?

e56K_lab What would your level of support be if you learned that each of the interim storage facilities for used nuclear fuel also would include a national research laboratory for studying ways to more safely and efficiently manage and transport nuclear materials?

Strongly Oppose						Strongly Support		
%	1	2	3	4	5	6	7	Mean
12 web-J	2	2	5	18	28	24	21	5.21
12 web-K	3	3	7	23	25	20	19	5.00

[12 web-J vs. K: $p = .0009$]

e57J_reuse What would your level of support be if you learned that each of the permanent repositories also would be designed to permit reprocessing used nuclear fuel for reuse in generating electricity?

e57K_reuse What would your level of support be if you learned that each of the interim storage facilities also would be designed to permit reprocessing used nuclear fuel for reuse in generating electricity?

Strongly Oppose						Strongly Support		
%	1	2	3	4	5	6	7	Mean
12 web-J	4	3	7	24	29	20	13	4.84
12 web-K	4	4	9	25	27	20	12	4.72

[12 web-J vs. 12 web-K: $p = .0735$]

e58J_comp What would your level of support be if you learned that the states and local communities hosting the permanent repositories for used nuclear fuel would receive (random: 10, 25, 100, 300) million dollars a year, paid for by revenues from nuclear power utilities, that could be used for public expenditures or to reduce state and local taxes?

e58K_comp What would your level of support be if you learned that the states and local communities hosting the interim storage facilities for used nuclear fuel would receive (random: 10, 25, 50, 300) million dollars a year, paid for by revenues from nuclear power utilities, that could be used for public expenditures or to reduce state and local taxes?

Strongly Oppose						**Strongly Support**		
%	**1**	**2**	**3**	**4**	**5**	**6**	**7**	**Mean**
12 web-J (10)	4	5	9	27	20	16	19	4.79
12 web-J (25)	4	3	6	23	29	23	12	4.88
12 web-J (100)	6	6	6	19	31	20	13	4.75
12 web-J (300)	5	4	6	23	26	21	15	4.81
12 web-K (10)	6	6	9	27	23	16	14	4.60
12 web-K (25)	5	4	7	21	29	20	13	4.77
12 web-K (50)	6	6	11	19	29	16	13	4.63
12 web-K (300)	8	4	7	21	24	18	17	4.71

[12 web-J (10) vs. J (300): $p = .8593$] [12 web-K (10) vs. K (300): $p = .4576$]

e59J_nmby1: What would your level of support be if you learned that one of these permanent repositories for used nuclear fuel is to be located in [insert state]?

e59K_nmby1: What would your level of support be if you learned that one of these interim storage facilities for used nuclear fuel is to be located in [insert state]?

Strongly Oppose						Strongly Support		
%	1	2	3	4	5	6	7	Mean
12 web-J	12	7	11	24	19	16	10	4.20
12 web-K	12	8	10	26	20	14	10	4.19

[12 web-J vs. K: $p = .9635$]

e60J_nmby2: What would your level of support be if you learned that one of these permanent repositories for used nuclear fuel is to be located (random: 50, 100, 300) miles from your principle residence?

e60K_nmby2: What would your level of support be if you learned that one of these interim storage facilities for used nuclear fuel is to be located (random: 50, 100, 300) miles from your principle residence?

Strongly Oppose						Strongly Support		
%	1	2	3	4	5	6	7	Mean
				2012 WEB-J				
50 miles	20	13	12	22	18	10	6	3.60
100 miles	17	10	7	21	22	14	9	4.00
300 miles	10	7	9	25	23	13	14	4.35
				2012 WEB-K				
50 miles	17	11	12	26	15	11	8	3.74
100 miles	19	7	11	22	23	12	7	3.87
300 miles	12	6	12	24	21	13	13	4.25

[12 web-J: 50 vs. 100: $p = .0075$; 50 vs. 300: $p < .0001$; 100 vs. 300: $p = .0127$]
[12 web-K: 50 vs. 100: $p = .3775$; 50 vs. 300: $p = .0004$; 100 vs. 300: $p = .0054$]

GROUP-J

There are at least two alternative approaches for choosing suitable sites for long-term storage and disposal of used nuclear fuel. ***In each approach, government regulators evaluate whether a site can safely contain nuclear materials for thousands of years using the same safety requirements.*** We briefly describe each approach below and ask you to rate your support or opposition.

GROUP-K

There are at least two alternative approaches for choosing suitable sites for interim storage facilities for used nuclear fuel. ***In each approach, government regulators evaluate whether a site can safely contain nuclear materials for up to a hundred years using the same safety requirements.*** We briefly describe each approach below and ask you to rate your support or opposition.

[e61J/K and e62J/K in Random Order]

***e61J_top Lead-in*:** In this option, Congress directs the federal government to identify two sites, one in the western U.S. and one in the east, that technical experts determine to be suitable for hosting nuclear repositories. Federal legislation is passed directing the states and communities where the sites are located to host a national repository for used nuclear fuel. Federal agencies work with the selected states and local communities to minimize negative economic, environmental, and social impacts while also creating hundreds of jobs and large investments. **This process places priority on technical experts first finding suitable sites, then working with the affected states and communities to meet their concerns.**

***e61K_top Lead-in*:** In this option, Congress directs the federal government to identify sites that technical experts determine to be suitable for hosting interim nuclear storage facilities. Federal legislation is passed directing the states and local communities where the sites are located to host an interim storage facility for used nuclear fuel. Federal agencies work with the selected states and local communities to minimize negative economic, environmental, and social impacts while also creating hundreds of jobs and large investments. **This process places priority on technical experts first finding suitable sites, then working with the affected states and communities to meet their concerns.**

e61_top_rate: On a continuous scale from one to seven, where one means you *strongly oppose* this policy process and seven means you *strongly support* it, how do you rate this site selection process?

Strongly Oppose						**Strongly Support**		
%	**1**	**2**	**3**	**4**	**5**	**6**	**7**	**Mean**
12 web-J (1)	4	5	9	33	26	15	8	4.49
12 web-J (2)	4	6	10	31	25	15	9	4.49
12 web-K (1)	5	4	12	30	26	14	9	4.46
12 web-K (2)	6	5	13	28	24	15	9	4.42

[12 web-J (1) vs. 12 web-K (1): $p = .6584$]

***e62J_botm Lead-in*:** In this option, Congress invites states and local communities to apply and compete to host a national repository for used nuclear fuel that will create hundreds of jobs and large investments. Federal agencies then work with qualified states and communities who want to compete, and the sites that are judged most suitable by technical experts are chosen to host a national nuclear repository. **This process places priority on first finding supportive host communities, then technical experts selecting the most suitable sites among them.**

***e62K_botm Lead-in*:** In this option, Congress invites states and local communities to apply and compete to host an interim storage facility for used nuclear fuel that will create hundreds of jobs and large investments. Federal agencies then work with qualified states and communities who want to compete, and the sites that are judged most suitable by technical experts are chosen to host interim nuclear storage facilities. **This process places priority on first finding supportive host communities, then technical experts selecting the most suitable sites among them.**

e62_botm_rate: On a continuous scale from one to seven, where one means you *strongly oppose* this policy process and seven means you *strongly support* it, how do you rate this site selection process?

Strongly Oppose						**Strongly Support**		
%	**1**	**2**	**3**	**4**	**5**	**6**	**7**	**Mean**
12 web-J (1)	4	5	9	31	26	14	11	4.53
12 web-J (2)	5	5	9	29	25	15	11	4.54
12 web-K (1)	5	4	10	32	26	14	10	4.53
12 web-K (2)	5	4	9	31	24	15	11	4.54

[12 web-J (1) vs. 12 web-K (1): $p = .9310$]

Now that you have recorded your level of support or opposition to each of these two site selection processes, we need you to rank them from the *most* preferred to the *least* preferred.

e63J/K_siterank Please use the drop-down boxes to assign a preference. When considering how to rank them, you may change the one to seven rating you previously assigned if you want to do so.
[NOTE: In above tables, (1) = first rating; (2) = second rating of same items.]

%	Least Preferred 1 (e61_top)	Most Preferred 2 (e62_botm)
12 web-J	49.4	50.6
12 web-K	49.4	50.6

Stop (Temporarily) Split J/K

Managing used nuclear fuel can be technically complex, and getting information you can trust is important. Please indicate your level of trust in information provided by science and engineering experts from each of the following organizations using a scale from zero to ten, where zero means *no trust* and ten means *complete trust*. [e64–e73 Randomized]

e64_NRC The U.S. Nuclear Regulatory Commission

No Trust									Complete Trust		
%	0	1	2	3	4	5	6	7	8	9	10	Mean
12 web	5	3	3	6	7	18	13	14	14	10	6	5.90

e65_EPA The U.S. Environmental Protection Agency

No Trust									Complete Trust		
%	0	1	2	3	4	5	6	7	8	9	10	Mean
12 web	7	3	4	4	7	14	13	14	14	10	9	5.94

e66_labs U.S. national laboratories for energy and security

No Trust									Complete Trust		
%	0	1	2	3	4	5	6	7	8	9	10	Mean
12 web	4	2	3	4	8	18	14	15	17	10	6	6.07

e67_NAS The National Academy of Sciences

No Trust									Complete Trust		
%	**0**	**1**	**2**	**3**	**4**	**5**	**6**	**7**	**8**	**9**	**10**	**Mean**
12 web	3	2	2	3	8	17	13	15	16	12	9	6.41

e68_state State regulatory agencies

No Trust									Complete Trust			
%	**0**	**1**	**2**	**3**	**4**	**5**	**6**	**7**	**8**	**9**	**10**	**Mean**
12 web	6	4	6	8	10	21	13	13	10	6	4	5.23

e69_NGO Environmental advocacy groups, such as the National Resources Defense Council or the Sierra Club

No Trust									Complete Trust			
%	**0**	**1**	**2**	**3**	**4**	**5**	**6**	**7**	**8**	**9**	**10**	**Mean**
12 web	8	4	5	6	9	17	12	11	13	8	7	5.52

e70_NEI The Nuclear Energy Institute, which represents the nuclear power industry

No Trust									Complete Trust			
%	**0**	**1**	**2**	**3**	**4**	**5**	**6**	**7**	**8**	**9**	**10**	**Mean**
12 web	8	5	6	8	11	17	11	12	10	7	5	5.15

e71_util Utility companies that own nuclear power plants

No Trust									Complete Trust			
%	**0**	**1**	**2**	**3**	**4**	**5**	**6**	**7**	**8**	**9**	**10**	**Mean**
12 web	11	7	8	10	11	17	12	9	7	4	3	4.40

e72_DOE The U.S. Department of Energy

No Trust									Complete Trust			
%	**0**	**1**	**2**	**3**	**4**	**5**	**6**	**7**	**8**	**9**	**10**	**Mean**
12 web	6	3	4	5	8	18	13	15	13	8	6	5.72

e73_fedcorp A private company chartered by the government and funded by fees from nuclear energy that is given responsibility for managing used nuclear fuel from U.S. nuclear power plants.

No Trust									Complete Trust		
%	0	1	2	3	4	5	6	7	8	9	10	Mean
12 web	11	7	7	9	10	19	11	9	8	5	3	4.54

Now we want to know more about impressions you may have about how these organizations are likely to assess risks associated with managing used nuclear fuel. Using a scale from one to seven, where one means the organization is likely to *downplay* risks, four means the organization is likely to *accurately assess* risks, and seven means the organization is likely to *exaggerate* risks, please rate your impressions of how each organization is likely to assess risks. [e64a–e73a Randomized]

e64a_NRC_rsk The U.S. Nuclear Regulatory Commission

Downplay Risks			Accurately Report Risks			Exaggerate Risks		
%	1	2	3	4	5	6	7	Mean
12 web	7	8	20	45	12	5	3	3.75

e65a_EPA_rsk The U.S. Environmental Protection Agency

Downplay Risks			Accurately Report Risks			Exaggerate Risks		
%	1	2	3	4	5	6	7	Mean
12 web	5	5	13	39	20	10	8	4.29

e66a_labs_rsk U.S. national laboratories for energy and security

Downplay Risks			Accurately Report Risks			Exaggerate Risks		
%	1	2	3	4	5	6	7	Mean
12 web	5	7	18	52	12	4	2	3.82

e67a_NAS_rsk The National Academy of Sciences

Downplay Risks			Accurately Report Risks			Exaggerate Risks		
%	**1**	**2**	**3**	**4**	**5**	**6**	**7**	**Mean**
12 web	3	3	10	59	16	5	3	4.10

e68a_state_rsk State regulatory agencies

Downplay Risks			Accurately Report Risks			Exaggerate Risks		
%	**1**	**2**	**3**	**4**	**5**	**6**	**7**	**Mean**
12 web	6	10	24	36	15	6	3	3.75

e69a_NGO_rsk Environmental advocacy groups, such as the National Resources Defense Council or the Sierra Club

Downplay Risks			Accurately Report Risks			Exaggerate Risks		
%	**1**	**2**	**3**	**4**	**5**	**6**	**7**	**Mean**
12 web	3	4	9	29	22	17	16	4.80

e70a_NEI_rsk The Nuclear Energy Institute, which represents the nuclear power industry

Downplay Risks			Accurately Report Risks			Exaggerate Risks		
%	**1**	**2**	**3**	**4**	**5**	**6**	**7**	**Mean**
12 web	16	16	23	30	9	4	2	3.19

e71a_util_rsk Utility companies that own nuclear power plants

Downplay Risks			Accurately Report Risks			Exaggerate Risks		
%	**1**	**2**	**3**	**4**	**5**	**6**	**7**	**Mean**
12 web	23	20	23	22	7	3	2	2.89

e72a_DOE_rsk The U.S. Department of Energy

Downplay Risks			Accurately Report Risks			Exaggerate Risks		
%	1	2	3	4	5	6	7	Mean
12 web	7	8	21	42	13	6	3	3.75

e73a_fedcorp A private company chartered by the government and funded by fees from nuclear energy that is given responsibility for managing used nuclear fuel from U.S. nuclear power plants.

Downplay Risks			Accurately Report Risks			Exaggerate Risks		
%	1	2	3	4	5	6	7	Mean
12 web	17	16	22	29	8	5	2	3.18

Split Design: Group-L (50%): No Mention of Shutdown of Japanese Nuclear Plants

As you may recall, a severe earthquake occurred on March 11, 2011 in the Pacific Ocean near Japan, creating large tidal waves that destroyed some Japanese coastal cities. Also damaged was the Fukushima nuclear power plant, which released radioactivity into the atmosphere and nearby portions of the sea. The earthquake and tidal wave killed thousands of people; the release of radiation at Fukushima is not known to have produced any deaths, but could contribute to future illnesses. We would like to know how the Japanese experience has influenced your confidence in U.S. nuclear power.

Split Design: Group-M (50%): Identifies Shutdown of Japanese Nuclear Plants

As you may recall, a severe earthquake occurred on March 11, 2011 in the Pacific Ocean near Japan, creating large tidal waves that destroyed some Japanese coastal cities. Also damaged was the Fukushima nuclear power plant, which released radioactivity into the atmosphere and nearby portions of the sea. The earthquake and tidal wave killed thousands of people; the release of radiation at Fukushima is not known to have produced any deaths, but could contribute to future illnesses. Currently, all Japanese nuclear power plants have been shut down, and Japan is trying to meet its electricity requirements without nuclear energy. We would like to know how the Japanese experience has influenced your confidence in U.S. nuclear power.

e74_Jpn On a scale from minus ten to plus ten, where minus ten means the Japanese experience has strongly reduced your support for U.S. nuclear power production, zero means the Japanese experience has had no effect on your support, and plus ten means the Japanese experience has strongly increased your support, how have recent events in Japan influenced your support for nuclear power production in the United States?

Strongly Reduced										**No Effect**										**Strongly Increased**	
	-10	**-9**	**-8**	**-7**	**-6**	**-5**	**-4**	**-3**	**-2**	**-1**	**0**	**1**	**2**	**3**	**4**	**5**	**6**	**7**	**8**	**9**	**10**	**Mean**
12 web L	8	2	2	3	4	4	3	6	6	7	39	1	2	2	1	2	1	1	2	1	2	– 1.56
12 web M	9	2	2	3	2	5	4	5	5	5	41	2	2	2	3	1	2	1	1	1	2	– 1.43

[12 web-L vs. M: $p = .5274$]

End Split L/M; Resume Split J/K

Split Design: Group-J (50%): Long-Term Repository

For purposes of this survey, assume that a small rural community that is about 100 miles from your primary residence in [insert state] has volunteered to be considered for hosting a permanent repository for used nuclear fuel.

Split Design: Group-K (50%): Interim Storage Facilities

For purposes of this survey, assume that a small rural community that is about 100 miles from your primary residence in [insert state] has volunteered to be considered for hosting an interim storage facility for used nuclear fuel.

Group-J and Group-K

Now we want you to consider the issue of "consent." The primary questions are how consent can be granted and how it can be withdrawn during the site selection process. The siting process involves numerous groups who have a stake in decision making (stakeholders). Among key stakeholders are

a. the host community and those who live near the proposed site
b. other residents of the host state
c. residents of bordering or nearby states
d. county and state governments, legislatures, and associated regulatory bodies
e. federal departments and agencies authorized to oversee the management of radioactive materials
f. federal environmental protection and regulatory agencies
g. nongovernmental organizations such as environmental groups
h. the U.S. Congress which oversees and helps fund nuclear materials management
i. and the nuclear energy industry whose utility companies generate electricity and produce used nuclear fuel

e74a_J_need On a scale where zero means *not at all important* and ten means *extremely important*, how important is it that "consent" must be granted by key stakeholders before siting a permanent nuclear repository for used nuclear fuel?

e74a_K_need On a scale where zero means *not at all important* and ten means *extremely important*, how important is it that "consent" must be granted by key stakeholders before siting an interim storage facility for used nuclear fuel?

Not At All Important									Extremely Important		
%	**0**	**1**	**2**	**3**	**4**	**5**	**6**	**7**	**8**	**9**	**10**	**Mean**
12 web-J	1	0	1	2	6	12	8	15	18	11	25	7.47
12 web-K	1	0	0	2	4	13	8	16	17	10	29	7.64

[12 web-J vs. K: $p = .0742$]

GROUP-J

Deciding what constitutes "consent" to build a permanent repository for used nuclear fuel and deciding which stakeholders should be involved are complex issues. The answers may vary depending on geographical, social, political, and other factors, so a Blue Ribbon Commission appointed by the President recommended that the issue of consent should be negotiated with the volunteer host community and state as part of the siting process.

GROUP-K

Deciding what constitutes "consent" to build an interim storage facility for used nuclear fuel and deciding which stakeholders should be involved are complex issues. The answers may vary depending on geographical, social, political, and other factors, so a Blue Ribbon Commission appointed by the President recommended that the issue of consent should be negotiated with the volunteer host community and state as part of the siting process.

e75J_consent / e75K_consent With which of the following broad approaches to determining consent do you most agree? [Arguments in Random Order]

1 "Consent" should involve a process where many different stakeholders must agree. Thus consent should require agreement by local elected officials, [insert state]'s governor, both of [insert state]'s U.S. senators, the U.S. congressperson representing the host community, and [insert state]'s environmental protection agencies. In

addition, a state-wide vote should be held that wins the support of a majority of citizens in [insert state].

2 "Consent" should involve a process where only those that are most affected must agree. Thus consent should require agreement by local elected officials and [insert state]'s governor. In addition, a vote should be held that wins the support of a majority of the residents in the local host community.

	More Inclusive	**Less Inclusive**
%	**1**	**2**
12 web-J	59	41
12 web-K	57	43

e76J_veto Please select all those on the following list that you think should be allowed to block/veto construction of the proposed permanent repository for used nuclear fuel in [insert state]:

e76K_veto Please select all those on the following list that you think should be allowed to block/veto construction of the proposed interim storage facility for used nuclear fuel in [insert state]:

	Permanent Repository	**Interim Storage**
%	**e76J**	**e76K**
1 - The Governor of [insert state]	56	54
2 – Either of the two U.S. senators from [insert state]	35	33
3 – The U.S. congressperson representing the district in which the host community is located	34	35
4 – The leaders of [insert state]'s legislature	29	28
5 – [insert state]'s environmental protection agency or its equivalent	47	49
6 – A majority of the citizens residing within 50 miles of the proposed facilities	67	69
7 – A majority of the voters of [insert state]	58	56
8 – The U.S. Nuclear Regulatory Commission	39	38
9 – The U.S. Environmental Protection Agency	40	41
10 – The U.S. Department of Energy	36	36
11 – Nongovernmental environmental interest groups in [insert state]	21	18

Group-J and Group-K

A related issue involves if and when consent might be withdrawn. The siting process will proceed in stages, and at some point a final decision to build or not to build the facility must be made. Each of these stages requires considerable investment of money and time. Each stage also provides more information for making a good decision. Generally, these stages include

GROUP-J

Step 1: The community or state volunteers to be a candidate to host a permanent repository for used nuclear fuel, and a technical evaluation of the site is begun. This evaluation may take several years to complete.

e77J_step1 Should the host state and local community be allowed to withdraw their consent during this stage?

	No	Yes
%	**0**	**1**
12 web-J	26	74

Step 2: Scientific evaluation of the suitability of the site for permanent storage and disposal of used nuclear fuel is completed.

e78J_step2 Should the host state and local community be allowed to withdraw their consent at this stage?

	No	Yes
%	**0**	**1**
12 web-J	26	74

Step 3: If the site is determined to be suitable, a license to construct a permanent repository for used nuclear fuel is submitted to U.S. regulatory agencies; the regulatory consideration may take several years to complete.

e79J_step3 Should the host state and local community be allowed to withdraw their consent during this stage?

	No	Yes
%	**0**	**1**
12 web-J	38	62

Step 4: If the license is provided, construction of a permanent repository for used nuclear fuel begins. Construction will take several years to complete.

e80J_step4 Should the host state and local community be allowed to withdraw their consent during this stage?

	No	**Yes**
%	**0**	**1**
12 web-J	60	40

Step 5: Construction is completed, and the permanent repository is prepared to receive used nuclear fuel.

e81J_step5 Should the host state and local community be allowed to withdraw their consent at this stage?

	No	**Yes**
%	**0**	**1**
12 web-J	70	30

GROUP-K

Step 1: The community or state volunteers to be a candidate to host an interim storage facility for used nuclear fuel, and a technical evaluation of the site is begun. This evaluation may take several years to complete.

e77K_step1 Should the host state and local community be allowed to withdraw their consent during this stage?

	No	**Yes**
%	**0**	**1**
12 web-K	25	75

[12 web-J vs. K: Chi Sq = 0.04; Fisher's exact $p = .8778$]

Step 2: Scientific evaluation of the suitability of the site for interim storage of used nuclear fuel is completed.

e78K_step2 Should the host state and local community be allowed to withdraw their consent at this stage?

	No	Yes
%	**0**	**1**
12 web-K	27	73

[12 web-J vs. K: Chi Sq = 0.26; Fisher's exact $p = .6145$]

Step 3: If the site is determined to be suitable, a license to construct an interim storage facility for used nuclear fuel is submitted to U.S. regulatory agencies; the regulatory consideration may take several years to complete.

e79K_step3 Should the host state and local community be allowed to withdraw their consent during this stage?

	No	Yes
%	**0**	**1**
12 web-K	34	66

[12 web-J vs. K: Chi Sq = 3.40; Fisher's exact $p = .0680$]

Step 4: If the license is provided, construction of an interim storage facility for used nuclear fuel begins. Construction will take several years to complete.

e80K_step4 Should the host state and local community be allowed to withdraw their consent during this stage?

	No	Yes
%	**0**	**1**
12 web-K	54	46

[12 web-J vs. K: Chi Sq = 6.76; Fisher's exact $p = .0099$]

Step 5: Construction is completed, and the interim storage facility is prepared to receive used nuclear fuel.

e81K_step5 Should the host state and local community be allowed to withdraw their consent at this stage?

	No	Yes
%	0	1
12 web-K	66	34

[12 web-J vs. K: Chi Sq = 2.63; Fisher's exact $p = .1146$]

End Split J/K

The next several questions are about your beliefs concerning a variety of issues.

e82_environ On a scale where zero means the natural environment is not at all threatened and ten means the natural environment is on the *brink of disaster*, how do you assess the current state of the natural environment?

Not At All Threatened									Brink of Disaster		
%	0	1	2	3	4	5	6	7	8	9	10	Mean
12 web	2	2	4	9	10	21	18	19	9	2	4	5.50

e83_doright On a scale from zero to ten, where zero means *none of the time* and ten means all of the time, how much of the time do you trust the government in Washington to do what is right for the American people?

None of the Time									All of the Time		
%	0	1	2	3	4	5	6	7	8	9	10	Mean
12 web	12	11	13	16	11	18	7	7	3	1	1	3.57

Now, please respond to each of the following statements using a scale from one to seven, where one means *strongly disagree* and seven means *strongly agree*. [e84–e95 Randomized]

e84_egal_1 What society needs is a fairness revolution to make the distribution of goods more equal.

Strongly Disagree						**Strongly Agree**		
%	**1**	**2**	**3**	**4**	**5**	**6**	**7**	**Mean**
12 web	13	8	10	23	19	14	13	4.19

e85_indiv1 Even if some people are at a disadvantage, it is best for society to let people succeed or fail on their own.

Strongly Disagree						**Strongly Agree**		
%	**1**	**2**	**3**	**4**	**5**	**6**	**7**	**Mean**
12 web	6	8	13	25	20	16	12	4.42

e86_hier1 The best way to get ahead in life is to work hard to do what you are told to do.

Strongly Disagree						**Strongly Agree**		
%	**1**	**2**	**3**	**4**	**5**	**6**	**7**	**Mean**
12 web	5	6	12	24	23	17	13	4.57

e87_fatal1 The most important things that take place in life happen by chance.

Strongly Disagree						**Strongly Agree**		
%	**1**	**2**	**3**	**4**	**5**	**6**	**7**	**Mean**
12 web	13	17	18	25	16	7	4	3.51

e88_egal2 Society works best if power is shared equally.

Strongly Disagree						**Strongly Agree**		
%	**1**	**2**	**3**	**4**	**5**	**6**	**7**	**Mean**
12 web	4	5	9	24	23	18	17	4.79

e89_indiv2 Even the disadvantaged should have to make their own way in the world.

Strongly Disagree						Strongly Agree		
%	1	2	3	4	5	6	7	Mean
12 web	6	9	15	27	21	13	9	4.25

e90_hier2 Society is in trouble because people do not obey those in authority.

Strongly Disagree						Strongly Agree		
%	1	2	3	4	5	6	7	Mean
12 web	10	12	16	24	20	11	9	4.01

e91_fatal2 No matter how hard we try, the course of our lives is largely determined by forces beyond our control.

Strongly Disagree						Strongly Agree		
%	1	2	3	4	5	6	7	Mean
12 web	9	14	14	21	19	13	10	4.05

e92_egal3 It is our responsibility to reduce differences in income between the rich and the poor.

Strongly Disagree						Strongly Agree		
%	1	2	3	4	5	6	7	Mean
12 web	13	11	11	21	18	14	12	4.09

e93_indiv3 We are all better off when we compete as individuals.

Strongly Disagree						Strongly Agree		
%	1	2	3	4	5	6	7	Mean
12 web	8	8	14	24	21	14	11	4.28

e94_hier3 Society would be much better off if we imposed strict and swift punishment on those who break the rules.

Strongly Disagree						**Strongly Agree**		
%	**1**	**2**	**3**	**4**	**5**	**6**	**7**	**Mean**
12 web	7	7	12	22	22	17	14	4.53

e95_fatal3 For the most part, succeeding in life is a matter of chance.

Strongly Disagree						**Strongly Agree**		
%	**1**	**2**	**3**	**4**	**5**	**6**	**7**	**Mean**
12 web	13	18	17	22	17	8	6	3.56

Different people rely on different sources of information about public issues. On average, approximately how many hours per week do you spend acquiring information on public issues from each of the following sources?

e96_srce1 Newspapers?

Trimmed Mean (50)	
12 Web	3.59

e97_srce2 Broadcast or cable television?

Trimmed Mean (50)	
12 Web	9.04

e98_srce3 The Internet, including news sources, blogs, discussion groups, etc.?

Trimmed Mean (50)	
12 Web	8.78

Please rate the degree to which each of the following four groups of statements describes your outlook on life, using a scale from zero to ten, where zero means *not at all* and ten means *completely*.

[e99_H_rate—e102_F_rate Random Order]

e99_H_rate: I am more comfortable when I know who is, and who is not, a part of my group, and loyalty to the group is important to me. I prefer to know who is in charge and to have clear rules and procedures; those who are in charge should punish those who break the rules. I like to have my responsibilities clearly defined, and I believe people should be rewarded based on the position they hold and their competence. Most of the time, I trust those with authority and expertise to do what is right for society.

Not at All										**Completely**		
%	**0**	**1**	**2**	**3**	**4**	**5**	**6**	**7**	**8**	**9**	**10**	**Mean**
12 web (1)	5	4	7	8	10	17	11	12	13	7	7	5.44
12 web (2)	5	5	7	8	10	16	11	12	12	7	6	5.41

[12 web (1) vs. 12 web (2): $p = .2646$]

e100_I_rate: Groups are not all that important to me. I prefer to make my own way in life without having to follow other peoples' rules. Rewards in life should be based on initiative, skill, and hard work, even if that results in inequality. I respect people based on what they do, not the positions or titles they hold. I like relationships that are based on negotiated "give and take," rather than on status. Everyone benefits when individuals are allowed to compete.

Not at All										**Completely**		
%	**0**	**1**	**2**	**3**	**4**	**5**	**6**	**7**	**8**	**9**	**10**	**Mean**
12 web (1)	3	2	4	5	9	16	11	11	14	10	14	6.32
12 web (2)	2	3	5	6	9	15	11	12	14	10	14	6.28

[12 web (1) vs. 12 web (2): $p = .2903$]

e101_E_rate: Much of society today is unfair and corrupt, and my most important contributions are made as a member of a group that promotes justice and equality. Within my group, everyone should play an equal role without differences in rank or authority. It is easy to lose track of what is important, so I have to keep a close eye on the actions of my group. It is not enough to provide equal opportunities; we also have to try to make outcomes more equal.

Not at All									Completely		
%	0	1	2	3	4	5	6	7	8	9	10	Mean
12 web (1)	9	5	7	8	10	19	10	10	9	6	7	5.03
12 web (2)	8	6	7	9	10	18	10	10	10	6	7	5.02

[12 web (1) vs. 12 web (2): $p = .6385$]

e102_F_rate: Life is unpredictable and I have little control. I have to live by lots of rules, but I don't get to make them. My fate in life is determined mostly by chance. I can't become a member of the groups that make most of the important decisions affecting me. Getting along in life is largely a matter of doing the best I can with what comes my way, so I focus on taking care of myself and the people closest to me.

Not at All									Completely		
%	0	1	2	3	4	5	6	7	8	9	10	Mean
12 web (1)	13	10	11	10	9	16	8	9	6	4	5	4.20
12 web (2)	11	10	12	10	9	15	8	8	7	4	5	4.22

[12 web (1) vs. 12 web (2): $p = .4556$]

Now that you have rated how well each of the four groups of statements describes your outlook, we need you to rank them from the one with which you *most* agree to the one with which you *least* agree. It is OK if you do not completely agree or completely disagree with any of the four groups of statements. (When considering how to rank them, you may change the zero to ten rating you previously assigned if you want to do so.)

[NOTE: In the above tables, (1) = first rating; (2) = second rating of same item.]

Please use the drop-down boxes to assign a number from four (*most* agree) to one (*least* agree) for each group of statements. You can use a ranking number only once, and you must assign a rank to each group of statements before you can advance to the next page.

[e99_H_rank—e102_F_rank Random Order]

***e99_H_rank*:** I am more comfortable when I know who is, and who is not, a part of my group, and loyalty to the group is important to me. I prefer to know who is in charge and to have clear rules and procedures; those who are in charge should punish those who break the rules. I like to have my responsibilities clearly defined, and I believe people should be rewarded based on the position they hold and their competence. Most of the time, I trust those with authority and expertise to do what is right for society.

	Least Agree			**Most Agree**	
	1	**2**	**3**	**4**	**Mean**
12 web	21	25	29	25	2.57

e100_I_rank: Groups are not all that important to me. I prefer to make my own way in life without having to follow other peoples' rules. Rewards in life should be based on initiative, skill, and hard work, even if that results in inequality. I respect people based on what they do, not the positions or titles they hold. I like relationships that are based on negotiated "give and take," rather than on status. Everyone benefits when individuals are allowed to compete.

	Least Agree			**Most Agree**	
	1	**2**	**3**	**4**	**Mean**
12 web	16	20	24	40	2.87

e101_E_rank: Much of society today is unfair and corrupt, and my most important contributions are made as a member of a group that promotes justice and equality. Within my group, everyone should play an equal role without differences in rank or authority. It is easy to lose track of what is important, so I have to keep a close eye on the actions of my group. It is not enough to provide equal opportunities; we also have to try to make outcomes more equal.

	Least Agree			**Most Agree**	
	1	**2**	**3**	**4**	**Mean**
12 web	28	26	25	21	2.39

e102_F_rank: Life is unpredictable and I have little control. I have to live by lots of rules, but I don't get to make them. My fate in life is determined mostly by chance. I can't become a member of the groups that make most of the important decisions affecting me. Getting along in life is largely a matter of doing the best I can with what comes my way, so I focus on taking care of myself and the people closest to me.

	Least Agree			Most Agree	
	1	**2**	**3**	**4**	**Mean**
12 web	34	29	21	15	2.17

e103_party With which political party do you most identify?

	Democratic	Republican	Independent	Other Party
%	**1**	**2**	**3**	**4**
12 web	42	26	31	1

e104_iden Do you completely, somewhat, or slightly identify with that political party?

	Slightly	Somewhat	Completely	
%	**1**	**2**	**3**	**Mean**
12 web	9	56	35	2.26

e105_ideol On a scale of political ideology, individuals can be arranged from strongly liberal to strongly conservative. Which of the following best describes your views? Would you say that you are:

	Strongly Liberal	Liberal	Slightly Liberal	Middle of the Road	Slightly Conserv.	Conserv.	Strongly Conserv.	
%	**1**	**2**	**3**	**4**	**5**	**6**	**7**	**Mean**
12 web	7	16	11	34	12	15	6	3.97

e106_race Which of the following best describes your race or ethnic background?

	American Indian	Asian	Black	Hispanic	White	Something Else
%	**1**	**2**	**3**	**4**	**5**	**6**
12 web	1	4	12	15	67	1

e107_city In what city is your primary residence located? [verbatim]

e108_inc Please indicate which of the following income categories approximates the total estimated annual income for your *household* for the year 2010.

	<$10K	$10–20K	$20–30K	$30–40K	$40–50K	$50–60K	$60–70K
%	1	2	3	4	5	6	7
12 web	8	11	12	12	10	12	8

	$70–80K	$80–90K	$90–100K	$100–110K	$110–120K	$120–130K	$130–140K
%	8	9	10	11	12	13	14
12 web	6	4	3	2	2	1	1

	$140–150K	$150–160K	$160–170K	$170–180K	$180–190K	$190–200K	>$200 K
%	15	16	17	18	19	20	21
12 web	1	1	0	1	1	1	2

%	Median
12 web	$40–50K

e109_web About how often do you access the Internet using a computer or some sort of a smartphone, like a Blackberry or iPhone??

	Never	< Once/ Month	Several Times/ Month	Once/ Week	Several Times/ Week	Once or Twice/ Day	Several Times/ Day
%	0	1	2	3	4	5	6
12 web	NA	5	2	2	7	18	67

e110_FB About how often do you use Facebook?

	Never	< Once/ Month	Several Times/ Month	Once/ Week	Several Times/ Week	Once or Twice/ Day	Several Times/ Day
%	0	1	2	3	4	5	6
12 web	21	9	6	8	11	18	26

e111_goog About how often do you use Google?

	Never	< Once/ Month	Several Times/ Month	Once/ Week	Several Times/ Week	Once or Twice/ Day	Several Times/ Day
%	**0**	**1**	**2**	**3**	**4**	**5**	**6**
12 web	6	7	7	7	19	19	35

e112_twit About how often do you use Twitter?

	Never	< Once/ Month	Several Times/ Month	Once/ Week	Several Times/ Week	Once or Twice/ Day	Several Times/ Day
%	**0**	**1**	**2**	**3**	**4**	**5**	**6**
12 web	69	10	3	5	5	4	5

e113_pnlfreq About how often do you answer surveys on the Internet?

	Never	< Once/ Month	Several Times/ Month	Once/ Week	Several Times/ Week	Once or Twice/ Day	Several Times/ Day
%	**0**	**1**	**2**	**3**	**4**	**5**	**6**
12 web	NA	7	10	11	27	18	28

Scholars have learned that information often influences the way in which people answer survey questions. With this in mind, we are interested in whether you are taking the time to read the text that precedes each question. So, in order to demonstrate that you have read this text, please ignore the question below and click on the blue dot.

e114_instruct Which of the following devices do you typically use to answer surveys on the Internet?

1 – a computer
2– A tablet (like an iPad)
3– A smart phone (like a Blackberry or iPhone)

	Did Not Click Blue Dot	Clicked Blue Dot
%	0	1
12 web	80	20

End Notes

[1] The strategy, subject to Congressional authorization, includes a pilot interim storage facility within 8 yr (focusing on fuel from shut-down reactors, consistent with a BRC recommendation); a larger, full-scale interim storage facility within 12 yr (both consistent with a suggested BRC timeframe of between 5 to 10 yr); and a geologic repository within 35 yr (DOE 2013: 2).

[2] For an overview of research programs and reports by CRCM

[3] We excluded sites that are still being considered. We excluded, for example, the proposed Yucca Mountain repository site because the Obama Administration's action to withdraw the license application to the NRC is still under legal review.

[4] The population of cases for this analysis was compiled using a three-step process. In the first step, we used a number of sources to compile a list of operational and decommissioned plants. For the US, our primary source was the NRC website. For international cases, our primary source of information on operational and decommissioned plants was the Power Reactor Information System (PRIS), which is developed and maintained by the International Atomic Energy Agency (IAEA). In the second step, we compiled a list of cancelled facilities using country reports produced the World Nuclear Association and the list of cancelled sites mentioned in a recent report by the Nuclear Energy Institute (NEI 2011). In the third step, we compressively searched the Internet, electronic newspaper archives, and academic literature (journal articles and books) for additional mentions of cancelled nuclear facilities.

[5] As noted above, our list of cases excludes those in Russia, China, North Korea, and the former Soviet Union. This is chiefly because the information that was available for these countries was heavily weighted towards successfully sited nuclear facilities, with little (if any) available for proposed and then cancelled cases. In order to avoid biasing our model results, it was therefore necessary to exclude all cases from these countries.

[6] The full description of the Polity database can be found on the Policy IV website, at: http://www.systemicpeace.org/polity/polity4.htm. The polity score measure has been revised periodically to reflect refinements in the theoretical understanding of authority structures. The initial theoretical development of the measure can be found in Eckstein (1975).

[7] For more on the Polity IV database and how it is organized, please refer to Appendix B.

[8] Indicator variables (also called dummy or binary variables) are binary measures in which a value of one indicates the presence of an attribute (college degree; male gender; member of a racial/ethnic minority; etc.) and zero indicates the absence of the attribute.

[9] "Wunderland Kalkar: Nuclear Power Plant Turned Amusement Park." 2011. *Amusing Planet.* http://www.amusingplanet.com/2011/06/wunderland-kalkar-nuclear-power-plant.html. Last accessed on October 10, 2012.

[10] We tested for alternative time lags following the major nuclear events, ranging from one year to ten year. The best model fit was obtained with the three year lag, and therefore we used the three year window to code the cases for proximty to a nuclear event.

[11] The expected value of the outcome (Y) for a logistic model for the Stage 1 data is as follows:

$$E\{Y\} = [1 + \exp(-\beta_0 - \beta_1 X_1 - \beta_2 X_2 - \beta_3 X_3 - \beta_4 X_4)]^{-1}.$$

[12] The occurrence of documented public opposition was fully explained by variation in the Polity score. For that reason, in order to prevent model co-linearity, the Polity score was not included as an explanatory variable in the direct effects model explaining siting outcomes.

[13]Access to the iPoll database for this study was through a subscription held by the University of Oklahoma.

[14]The WCALC program is available online at: www.unc.edu/ jstimson/wcalc.pdf

[15] For example, in December 1976, a Harris polls asked 1,459 respondents; "Do you favor or oppose the building of more nuclear power plants in the United States?" 54% favored and 34% opposed, therefore the ratio of support to opposition was 0.61.

[16] We are not able to vary estimators of the length of the effect of Fukushima given that we only have 1 year of data following that event.

[17] The perceived public *benefits* of nuclear energy, which have not been measured with sufficient regularity in polling to permit the kind of analysis utilized here, are also important (Herron, Jenkins-Smith and Silva 2011). In other research, and as reported in Section 5 of this report, we have observed that the Fukushima event reduced the perceived benefits of nuclear energy as a reliable and secure source of electricity.

[18] For some of the classic work on the role of public opinion shaping policy, see Downs (1957) and Schattschneider (1960). For research demonstrating that policy change tends to follow changes in public opinion, see Page and Shapiro (1983).

[19] For more on this phenomenon, refer to May, Workman, and Jones (2008).

[20] This study was collaboratively conceived and conducted with funding from SNL for survey design and analysis, and funding from the University of Oklahoma for survey data collection.

[21] Indicator variables (also called dummy or binary variables) are binary measures in which a value of one indicates the presence of an attribute (college degree; male gender; member of a racial/ethnic minority; etc.) and zero indicates the absence of the attribute.

[22] Mary Douglas helped pioneer cultural theory by introducing a grid/group typology identifying four primary classifications: egalitarians, individualists, hierarchists, and fatalists (Douglas 1970). Subsequently the typology was applied to risk analysis (Douglas and Wildavsky 1982). Other important contributors to the study of political culture include Thompson and Wildavsky (1982), and Wildavsky and Dake (1990). For a quantitative test of cultural theory hypotheses, see Jenkins-Smith and Smith (1994).

[23] A split design was included to experimentally test the effect of informing respondents that, at the time of the survey, that "Currently, all Japanese nuclear power plants have been shut down, and Japan is trying to meet its electricity requirements without nuclear energy." There was no statistically significant effect of including the additional information.

[24] Some members of the public may understand (through direct experience from home construction) that the construction process involves a number of steps that includes seeking several permits and, for large construction, exploratory drill holes for foundations etc. However, because seeking a construction authorization was a major step for Yucca Mountain, those intimately involved with the YMP have commented that "halted construction" in the question would give the impression to the public that the NRC had issued a license for construction of the surface handling facilities and disposal drifts. However, the general public is not that familiar with the specific steps of repository construction and operation. A full explanation that construction of 5 miles of tunnel and numerous alcoves was not construction of the repository but rather construction of an exploratory studies facility that would become part of repository, and that DOE chose to withdraw the application seeking authorization to convert the exploratory studies facility into part of the repository and construct additional disposal drifts and surface handling facilities would easily distract the reader from the intent of the question.

[25] An alternative description of this option is where the waste disposal function is not separated from other functions related to fuel fabrication and reprocessing.

[26] One can also consider combining storage/disposal facility with non-fuel cycle facility options – such as national parks or hospitals – but that is beyond the scope of this study (e.g., see OECD-NEA 2007).

[27] Interestingly, varying the size of the compensation offered to hosting states and local communities (ranging from $10 million to $300 million per year) had no effect on support for the program. Note that this finding does not imply that residents of potential host communities will be unresponsive to compensation; the question refers to how compensation to a hypothetical host state/community would influence overall support for a national siting policy. Understanding the effect of compensation on residents of a candidate host community would require a different kind of analysis (e.g., Jenkins-Smith and Kunreuther 2001).

[28] For those participants who allowed us to record their residential location, we estimated proximity using the most precise of three sources of location data. For those who used equipment to take the survey that afforded exact latitude and longitude, we recorded precise location. If that kind of equipment was not used, we estimated location based on the respondent's Internet Protocol (IP) address. For all others, we estimated distance based on the centroid of the respondent's zip code area.

In: Nuclear Energy, Facility Siting ...
Editor: Francis Crowson
ISBN: 978-1-62948-787-8

Chapter 2

STRATEGY FOR THE MANAGEMENT AND DISPOSAL OF USED NUCLEAR FUEL AND HIGH-LEVEL RADIOACTIVE WASTE*

U.S. Department of Energy

In 2010, I chartered the *Blue Ribbon Commission on America's Nuclear Future* ("BRC" or "Commission") to conduct a comprehensive review and

* This is an edited, reformatted and augmented version of the U.S. Department of Energy publication, dated January 2013.

recommend a plan of action for the management and disposal of the nation's used nuclear fuel and high-level radioactive waste, also referred to as the back-end of the nuclear fuel cycle. Representative Lee Hamilton and General Brent Scowcroft, two distinguished individuals with decades of public service and governing experience, co-chaired the Commission and led a panel of leading scientists, nuclear energy experts, industry leaders, and former elected officials Nuclear power is an integral part of our "all-of-the-above" energy strategy. It provides twenty percent of our nation's electricity supply, and the Administration is promoting the safe use of nuclear power through support for new nuclear power plants incorporating state-of-the-art passive safety features as well as a cost-shared program providing technical support for licensing new small reactor designs. Nuclear energy is an important contributor to our nation's energy security, and promotes clean-energy jobs. Nuclear energy production also provides important environmental benefits by producing little carbon dioxide or conventional air pollutant emissions.

An unfailing commitment to protect public health and safety, security, and the environment is essential to ensuring that nuclear power remains part of our diversified clean-energy portfolio. As part of that commitment, safe, long-term management and disposal of used nuclear fuel and high-level radioactive waste must remain a national priority.

Beyond sustaining an important domestic energy source, progress on a disposal solution can also support the clean-up of those sites that hosted production of defense nuclear materials during the Cold War, and help advance key national-security and non-proliferation objectives. More than 40 percent of the Navy's surface and submarine combatant fleet, for example, is now nuclear-powered. The used nuclear fuel it generates likewise requires a permanent disposal solution.

Since the end of the Cold War, significant quantities of weapons-capable plutonium and highly enriched uranium have become surplus to our national security needs. Some of these nuclear materials will be modified so they can be used in reactors as fuel, but then will be destined for a repository.

Finally, global demand for nuclear energy continues to grow, with commensurate risks in terms of safety, weapons proliferation, and terrorism if this growth occurs outside a vigorous safety and security framework. America's ability to influence the mitigation of these risks is strengthened when we demonstrate the commitment and ability to perform here at home.

For nearly two years, the Commission conducted a comprehensive review and ultimately made recommendations for addressing one of our nation's most intractable challenges. Its work provides a strong foundation for development of a new strategy to manage used nuclear fuel and high-level radioactive waste. We will work with Congress to build a new national program based on this foundation.

INTRODUCTION AND SUMMARY

The *Strategy for the Management and Disposal of Used Nuclear Fuel and High-Level Radioactive Waste* is a framework for moving toward a sustainable program to deploy an integrated system capable of transporting, storing, and disposing of used nuclear fuel[1] and high-level radioactive waste from civilian nuclear power generation, defense, national security and other activities.

The Strategy addresses several important needs. First, it serves as a statement of Administration policy regarding the importance of addressing the disposition of used nuclear fuel and high-level radioactive waste; it lays out the overall design of a system to address that issue; and it outlines the reforms needed to implement such a system. Second, it presents the Administration's response to the final report and recommendations made by the *Blue Ribbon Commission on America's Nuclear Future* ("BRC"). It also responds to direction in the Joint Explanatory Statement accompanying the Consolidated Appropriations Act, 2012, to develop a strategy for the management of used nuclear fuel and nuclear waste in response to the BRC's recommendations. Third, this strategy represents an initial basis for discussions among the Administration, Congress and other stakeholders on a sustainable path forward for disposal of nuclear waste.

The Administration endorses the key principles that underpin the BRC's recommendations. The BRC's report and recommendations provide a starting point for this Strategy, which translates many of the BRC's principles into an actionable framework within which the Administration and Congress can build a national program for the management and disposal of the nation's used nuclear fuel and high-level radioactive waste.[2] The BRC report and the Strategy build on the body of physical and social science work completed during the prior decades and benefit from the lessons learned not only from our nation's experiences, but also from those of other countries.

This Strategy includes a phased, adaptive, and consent-based approach to siting and implementing a comprehensive management and disposal system. At its core, this Strategy endorses a waste management system containing a pilot interim storage facility; a larger, full-scale interim storage facility; and a geologic repository in a timeframe that demonstrates the federal commitment to addressing the nuclear waste issue, builds capability to implement a program to meet that commitment, and prioritizes the acceptance of fuel from shut-down reactors. A consent-based siting process could result in more than one storage facility and/or repository, depending on the outcome of discussions with host communities; the Nuclear Waste Policy Act of 1982 (NWPA) envisaged the need for multiple repositories as a matter of equity between regions of the country. As a starting place, this Strategy is focused on just one of each facility.

With the appropriate authorizations from Congress, the Administration currently plans to implement a program over the next 10 years that:

- Sites, designs and licenses, constructs and begins operations of a pilot interim storage facility by 2021 with an initial focus on accepting used nuclear fuel from shut-down reactor sites;
- Advances toward the siting and licensing of a larger interim storage facility to be available by 2025 that will have sufficient capacity to provide flexibility in the waste management system and allows for acceptance of enough used nuclear fuel to reduce expected government liabilities; and
- Makes demonstrable progress on the siting and characterization of repository sites to facilitate the availability of a geologic repository by 2048.

Full implementation of this program will require legislation to enable the timely deployment of the system elements noted above. Legislation should also include the requirements for consent-based siting; a reformed funding approach that provides sufficient and timely resources; and the establishment of a new organization to implement the program, the structure of which should balance greater autonomy with the need for continued Executive and Legislative branch oversight. The Administration looks forward to engaging Congress on comprehensive legislation to move forward on this important national responsibility.

In the meantime, the Administration, through the Department of Energy (DOE), is undertaking activities within existing Congressional authorization to plan for the eventual transportation, storage, and disposal of used nuclear fuel. Activities range from examining waste management system design concepts, to developing plans for consent-based siting processes, to conducting research and development on the suitability of various geologies for a repository. These activities are designed to not limit the options of either the Administration or Congress and could be transferred to the new waste management and disposal organization when it is established.

BACKGROUND

The NWPA established a broad policy framework for the permanent disposal of used nuclear fuel and high-level radioactive waste derived from nuclear power generation. The NWPA authorized the government to enter into contracts with reactor operators – the generators and current owners of used nuclear fuel – providing that, in exchange for the payment of fees, the government would assume responsibility for permanent disposal. The fees were to ensure that the reactor owners and power generators pay the full cost of the disposal of their used nuclear fuel and high-level radioactive waste.

The federal government did not meet its contractual obligation to begin accepting used nuclear fuel by 1998. As a result of litigation by contract holders, the government was found in partial breach of contract, and is now liable for damages to some utilities to cover the costs of on-site, at-reactor storage.

Currently more than 68,000 metric tons heavy metal (MTHM) of used nuclear fuel are stored at 72 commercial power plants around the country with approximately 2,000 MTHM added to that amount every year. The sooner that legislation enables progress on implementing this Strategy, the lower the ultimate cost will be to the taxpayers. This document outlines a strategy that is intended to limit, and then end, liability costs by making it possible for the government to begin performing on its contractual obligations.

The NWPA specified a process for evaluating sites for a repository. The Administration concurs with the conclusion of the BRC that a fundamental flaw of the 1987 amendments to the NWPA was the imposition of a site for characterization, rather than directing a siting process that is, as the BRC recommends, "explicitly adaptive, staged, and consent-based..." In practical terms, this means encouraging communities to volunteer to be considered to host a nuclear waste management facility while also allowing for the waste management organization to approach communities that it believes can meet the siting requirements. Under such an arrangement, communities could volunteer to provide a consolidated interim storage facility and/or a repository in expectation of the economic activity that would result from the siting, construction, and operation of such a facility in their communities.

In addition to commercial used nuclear fuel, high-level radioactive wastes that are the by-products of the production of the nation's nuclear weapons and used fuel from the Navy's nuclear powered combat vessels also require a defined disposal path. These wastes are currently stored at sites in Idaho, South Carolina, and Washington. Also, significant quantities of weapons-capable plutonium and highly enriched uranium have become surplus to our national security needs, and in some form will be destined for disposal in a repository.

STRATEGY ELEMENTS

This Strategy provides a basis for the Administration to work with Congress to design and implement a program to meet the government's obligation to take title to and permanently dispose of used nuclear fuel and high-level radioactive waste.

It also provides near-term steps to be implemented by DOE pending enactment of new legislation. The key elements of this Strategy are captured in Figure 1.

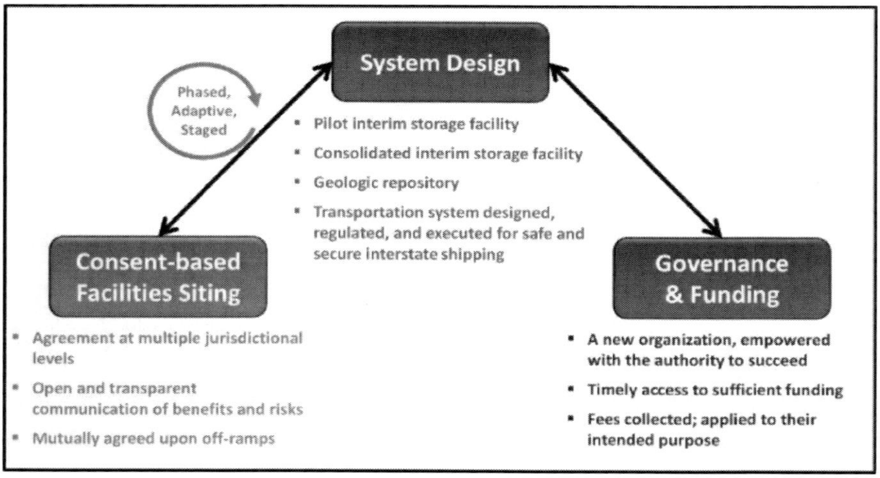

Figure 1. Key Strategy Elements.

System Design

The Administration supports an approach to system design that integrates consent-based siting principles and makes progress in demonstrating the federal commitment to addressing used nuclear fuel and high-level radioactive waste disposal, including building the capability to begin executing that commitment within the next 10 years. The Administration supports a nuclear waste management system with the following elements:

- A pilot interim storage facility with limited capacity capable of accepting used nuclear fuel and high-level radioactive waste and initially focused on serving shut-down reactor sites;
- A larger, consolidated interim storage facility, potentially co-located with the pilot facility and/or with a geologic repository, that provides the needed flexibility in the waste management system and allows for important near-term progress in implementing the federal commitment; and
- A permanent geologic repository for the disposal of used nuclear fuel and high-level radioactive waste.

The objective is to implement a flexible waste management system incrementally in order to ensure safe and secure operations, gain trust among stakeholders, and adapt operations based on lessons learned. As will be addressed in the following section on implementation, the Administration agrees with the Blue Ribbon Commission that a consent-based siting process offers the promise of sustainable decisions for both storage and disposal facilities. Figure 2 below portrays a set of possible pathways to developing system facilities and capabilities.

This system would initially be focused on acceptance of used nuclear fuel from shut-down reactors; such fuel provides an opportunity to build waste handling capability as well as to relieve surrounding communities and utility contract holders of the burdens associated with long-term storage of used nuclear fuel at a shut-down reactor. Following these initial efforts, capacity will be developed to enable the acceptance and transportation of used nuclear fuel at rates greater than that at which utilities are currently discharging it in order to gradually work off the current inventory. The Administration remains committed to addressing the Cold War legacy; and, in addition to ongoing efforts, will consider transportation and interim storage of government-owned used nuclear fuel and high-level radioactive waste at interim storage facilities.

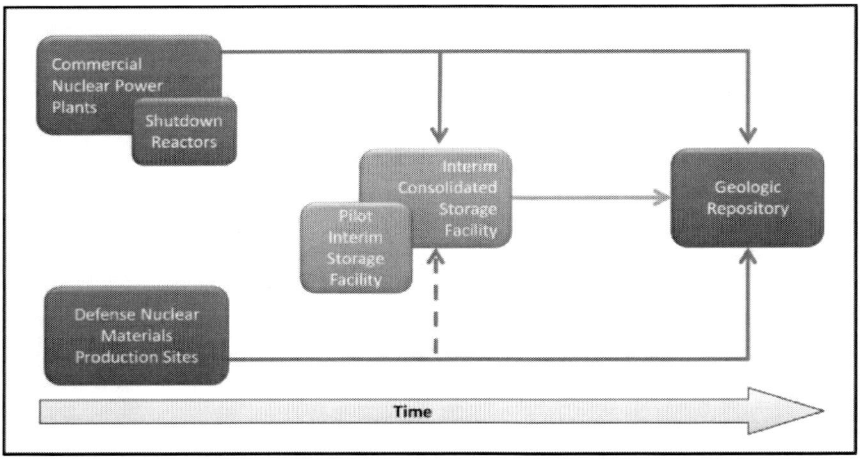

Figure 2. Possible system pathways.

Interim Storage

The BRC recommended that "one or more consolidated (interim) storage facilities be developed to start the orderly transfer of used nuclear fuel from reactor sites to safe and secure centralized facilities independent of the schedule for operating a permanent repository." The Administration agrees that interim storage should be included as a critical element in the waste management system and has several benefits, including flexibility in system planning and execution and the opportunity to move expeditiously to fulfill government contractual responsibilities.

The Administration also agrees with the BRC that a linkage between opening an interim storage facility and progress toward a repository is important so that states and communities that consent to hosting a consolidated interim storage facility do not face the prospect of a *de facto* permanent facility without consent. However, this linkage should not be such that it overly restricts forward movement on a pilot or larger storage facility that could make progress against the waste management mission. The NWPA currently constrains the development of a storage facility by limiting the start of construction of such a facility until after the Nuclear Regulatory Commission (NRC) has issued a license for construction of a repository. This restriction has effectively eliminated the possibility of having an interim storage facility as an integral component of a waste management system.

Consistent with legislation recently under consideration in Congress, the Administration supports the development of a pilot interim storage facility with an initial focus on accepting used nuclear fuel from shut-down reactor sites. Acceptance of used nuclear fuel from shut-down reactors provides a unique opportunity to build and demonstrate the capability to safely transport and store used nuclear fuel, and therefore to make progress on demonstrating the federal commitment to addressing the used nuclear fuel issue. A pilot would also build trust among stakeholders with regard to the consent-based siting process and commitments made with a host community for the facility itself, with jurisdictions along transportation routes, and with communities currently hosting at-reactor storage facilities if enabled by appropriate legislation. The Administration would plan to undertake activities necessary to enable the commencement of operations at this facility in 2021, including conducting a consent-based siting process with interested parties, undertaking the requisite analyses associated with siting such a facility, and initiating engineering and design activities as warranted. Full execution of this plan depends on enactment of revised legislative authority.

Beyond a pilot-scale facility, the Administration supports the development of a larger consolidated interim storage facility with greater capacity and capabilities that will provide flexibility in operation of the transportation system and disposal facilities. In addition, a larger-scale facility could take possession of sufficient quantities of used nuclear fuel to make progress on the reduction of long-term financial liabilities. Depending on the outcome of a consent-based process, this facility could have a capacity of 20,000 MTHM or greater, and could be co-located with the pilot facility or the eventual geologic repository. In the context of the overall waste management system, the Administration supports the goal of siting, designing, licensing, constructing and commencing operations at a consolidated interim storage facility by 2025.

In addition to commercial used nuclear fuel, pilot-scale and larger interim storage facilities could provide similar benefits for government-owned and managed used nuclear fuel and high-level radioactive waste, such as demonstration of capability and flexibility in system operations. Therefore, the feasibility of accepting these wastes at interim storage facilities will be considered.

Transportation

The BRC found that existing standards and regulations for the transportation of used nuclear fuel and high-level radioactive waste administered by DOE, NRC, the U.S. Department of Transportation, and state, local, and tribal governments are proven and functioning well. Consistent with the recommendations of the BRC on this issue, the Administration is moving ahead with initial planning for engagement and technical assistance for transportation operations for state and local governments.

As described in the Ongoing Activities section of this document, the Department is proceeding with planning activities for the development of transportation capabilities and storage facilities to facilitate the acceptance of used nuclear fuel at a pilot interim storage facility within the next 10 years and later at a larger consolidated interim storage facility. The Administration will undertake the transportation planning and acquisition activities necessary to initiate this process with the intent to transfer them to a separate organizational entity if and when it is authorized by Congress and in operation. Outreach and communication, route analysis, and emergency response planning activities consistent with existing NWPA requirements would be conducted during this time. The Administration agrees with the BRC that the relationships and processes built with other federal agencies, state agencies, and local governments to support logistics of shipments to the Waste Isolation Pilot Plant (WIPP) have been successful and the infrastructure and lessons learned from this experience will be utilized moving forward.

Geologic Disposal

There is international consensus that geologic repositories represent the best known method for permanently disposing of used nuclear fuel and high-level radioactive waste, without putting a burden of continued care on future generations. The BRC recommended that the U.S. undertake "an integrated nuclear waste management program that leads to the timely development of one or more permanent deep geologic facilities for the safe disposal of used fuel and high-level nuclear waste." The Administration agrees that the development of geologic disposal capacity is currently the most cost-effective way of permanently disposing of used nuclear fuel and high-level radioactive waste while minimizing the burden on future generations. As noted by the BRC, the linkage between storage and disposal is critical to maintaining confidence in the overall system. Therefore, efforts on implementing storage capabilities within the next 10 years will be accompanied by actions to engage in a consent-based siting process and begin to conduct preliminary site investigations for a geologic repository. The Administration's goal is to have a repository sited by 2026; the site characterized, and the repository designed and licensed by 2042; and the repository constructed and its operations started by 2048. Consistent with this effort, the Administration understands the need for the Environmental Protection Agency to develop a set of generic, non-site-specific, repository safety standards to gain public confidence that any future repository will protect public health and the environment. This will be an important early step in any repository siting effort.

The ability to retrieve used nuclear fuel and high-level radioactive waste from a geologic repository for safety purposes or future reuse has been a subject of repository design debate for many years. A recently completed technical review by Oak Ridge National Laboratory found that approximately 98 percent of the total current inventory of commercial used nuclear fuel by mass can proceed to permanent disposal without the need to ensure post-closure recovery for reuse based on consideration of the viability of economic recovery of nuclear materials, research and development (R&D) needs, time frames in which recycling might be deployed, the wide diversity of types of used nuclear fuel from past operations, and possible uses to support national security interests.[3]

This assessment does not preclude any decision about future fuel cycle options, but does indicate that retrievability it is not necessary for purposes of future reuse.

Disposal of defense wastes alongside commercial wastes is the current policy in accordance with the 1985 decision to use a single repository for both commercial and defense wastes. The issue of "commingling" of wastes in a repository will be the subject of analysis moving forward.

Advanced Fuel Cycles

The BRC concluded that "it is premature at this point for the United States to commit irreversibly to any particular fuel cycle as a matter of government policy..." and pointed out that "it is... very likely that disposal will be needed to safely manage at least some portion of the existing commercial [used nuclear fuel] inventory." Even if a closed fuel cycle were to be adopted in the future, permanent geologic disposal will still be required for residual high-level radioactive waste. Cost, nonproliferation, national security, environmental concerns, and technology limitations are some of the concerns that would need to be addressed before any future decision to close the U.S. fuel cycle through the use of recycling would be made. These factors reinforce the likelihood that the once-through fuel cycle will continue at least for the next few decades.

Nevertheless, consistent with past practice and the BRC's recommendations, DOE will continue to conduct research on advanced fuel cycles to inform decisions on new technologies that may contribute to meeting the nation's future energy demands while supporting non-proliferation and used nuclear fuel and high-level radioactive waste management objectives.

International Cooperation

International cooperation has been a cornerstone of both U.S. fuel cycle R&D efforts as well as actions to reduce the global proliferation of nuclear materials.

Recently, several countries, led by the U.S. and others, have come together to establish frameworks within which multi-national fuel cycle facilities could enable wider access to the benefits of nuclear power while reducing proliferation risks.

The BRC recommended that the U.S. develop the capability "to accept used fuel from foreign commercial reactors, in cases where the President would choose to authorize such imports for reasons of U.S. national security." The focus of the present Strategy is on a clear path for the safe and permanent disposal of U.S. used nuclear fuel and high-level radioactive waste; however, the Administration will continue to evaluate the BRC's recommendation and will discuss with Congress the pros and cons of including it in the new waste disposal program.

Implementation

Critical elements for successful implementation of this Strategy include the establishment of a consent-based siting process, a new organization to execute the waste management mission and implementation of a process for long-term, stable funding. The design of both the new organization and the funding source should strike an appropriate balance between independence of the new organization and the need for oversight by Congress and the Executive branch.

Consent-Based Siting

The BRC recommends a siting process that is consent-based, transparent, phased, adaptive, standards- and science-based, and governed by legally-binding agreements between the federal government and host jurisdictions. Indeed, promising experiences in other countries indicate that a consent-based process, developed through engagement with states, tribes, local governments, key stakeholders, and the public, offers a greater probability of success than a top down approach to siting. One of the consequences of a consent-based siting process could be the need to have more than one storage facility and/or repository. Multiple communities with differing interests and strengths may propose options leading to system configurations that involve multiple facilities. However, this Strategy focuses on one pilot storage, consolidated interim storage, and repository. The BRC offered the view that "a good gauge of consent would be the willingness of the host [jurisdictions] to enter into legally binding agreements...that can protect the interests of their citizens." Defining consent, deciding how that consent is codified, and determining whether or how it is ratified by Congress are critical first steps toward siting the storage facilities and repository discussed above. As such, they are among the near-term activities to be undertaken by the Administration in consultation with Congress and others. Legislation recently under consideration by Congress includes requirements for consent at multiple levels, including Congressional ratification. The Department is currently gathering information from the siting of nuclear facilities in the U.S. and elsewhere in order to better understand critical success factors in these efforts and to facilitate the development of a future siting process for a repository and storage facilities.

This Strategy endorses the proposition that prospective host jurisdictions must be recognized as partners. Public trust and confidence is a prerequisite to the success of the overall effort, as is a program that remains stable over many decades; therefore, public perceptions must be addressed regarding the program's ability to transport, store, and dispose of used nuclear fuel and high-level radioactive waste in a manner that is protective of the public's health, safety, and security and protective of the environment.

Management and Disposal Organization

A new waste management and disposal organization (MDO) is needed to provide the stability, focus, and credibility to build public trust and confidence.

Managing waste and used fuel is a governmental responsibility and there are multiple possible structures for this new organization. The MDO would be charged with the management and disposal of commercial used nuclear fuel and the associated interface with the utilities. The government will continue to manage its own high-level radioactive waste and used nuclear fuel until it is transferred to an MDO for storage and/or disposal. The BRC recommended the establishment of new, single-purpose organization "to provide the stability, focus, and credibility that are essential to get the waste program back on track." The BRC recommended a specific model in a congressionally-chartered federal corporation. The Administration agrees that a new organizational entity is needed and believes that there are several viable organizational models that can possess the critical attributes described below.

As part of the development of this Strategy, the Department of Energy commissioned work by the RAND Corporation to examine organizational alternatives for addressing used nuclear fuel and high-level radioactive wastes.[4] RAND assessed lessons learned from the history of the previous DOE organization and analyzed alternative organizational models currently in use both in and out of government. The study's authors concluded that a federal government corporation and an independent government agency are two promising models for a new organization to manage and dispose of used nuclear fuel and high-level radioactive waste, as both models can achieve the critical attributes of accountability, transparent decision-making, autonomy, a public interest mission, and organizational stability. The study also examined the attributes of federally-chartered private corporations and determined that this model is not a good option because obligations to stockholders and the profit motive could result in weakened public accountability and poor political credibility. The RAND study noted that "The success of any future MDO will be driven by many factors and unforeseen circumstances. The organizational form is only one of these factors and perhaps not even the most important one." Rather, of key importance is the flexibility the U.S. government has in crafting a new organization and the specific characteristics with which that organization is endowed.

Whatever form the new organization takes, organizational stability, leadership continuity, oversight and accountability, and public credibility are critical attributes for future success. The Administration will work with Congress to ensure that the MDO authorization provides adequate authority and leadership to execute its mission, with appropriate oversight and controls. Pending enactment of new legislation to establish the MDO, DOE's existing offices retain responsibility to maintain progress in implementing this

Strategy. Once the MDO is established, the Administration will carefully evaluate the appropriate activities to be transferred. DOE will take necessary steps to advance the program while taking every precaution to avoid compromising the later ability of the newly established MDO to succeed.

In addition, the mission of the MDO will need to be carefully defined. For example, funding made available to the MDO should be used only for the management and disposal of radioactive waste. While this could include the management and disposal of waste resulting from the processing of defense materials, the MDO itself should not be authorized to perform research on, fund or conduct activities to reprocess or recycle used nuclear fuel. These limitations on the MDO mission are consistent with the recommendations of the BRC.

Funding

With regard to funding, the BRC noted that "...the success of a revitalized nuclear waste management program will depend on making the revenues generated by the nuclear waste fee and the balance in the NWF available when needed and in the amounts needed to implement the program." The Administration agrees that providing adequate and timely funding is critical to the success of the nuclear waste mission.

The NWPA established a self-financing mechanism for the nation's commercial nuclear material management system. Congress intended at the time to ensure a stable, ongoing source of funding for the program and also one that would not burden taxpayers. Under the NWPA, the government currently assesses utilities a fee equal to one mill ($0.001) for each kilowatt-hour of electricity sold from nuclear power plants in exchange for agreeing to accept and permanently dispose of utilities' used nuclear fuel. Fees collected total approximately $750 million per year. This fee income is credited to the Nuclear Waste Fund (NWF, or the "Fund"), a fund held in the U.S. Treasury in which monies in excess of appropriations are invested in non-marketable Treasury securities, and the interest earnings are credited to the Fund. The current balance of the Fund is estimated at $28 billion.

Subsequent to passage of the NWPA, a series of broader budgeting acts passed by Congress have had the effect of disconnecting the revenues from the expenditures necessary for a waste disposal solution. All NWF spending is subject to annual appropriations and is required to compete with other priorities within budget caps imposed on all government discretionary spending, while continued collection of the full amount of fees is credited on the mandatory side of the budget as offsetting receipts. As a result, even though the intent of the NWPA was to make the balances of the NWF available when needed to cover the government's cost to dispose of the used nuclear fuel, there is a disconnect that makes access to funding difficult.

Moving forward, the key challenge is to ensure that past and future fee receipts and accrued interest are made available to meet mission requirements in a timely and dependable manner. To achieve this goal, reform of the current funding arrangement is necessary and should consist of the following elements: ongoing discretionary appropriations, access to annual fee collections provided in legislation either through their reclassification from mandatory to discretionary or as a direct mandatory appropriation of the fees, and eventual access to the balance or "corpus" of the NWF.

First, future funding arrangements should include a role for the Appropriations Committees of Congress through ongoing discretionary appropriations, funded within the discretionary spending limits. Ongoing engagement with the Appropriations committees ensures annual oversight and increases the likelihood of a sustained Congressional commitment to the nuclear waste mission. Annual appropriations could be used to fund expenses that are regular and recurring, such as program management costs, including administrative expenses, salaries and benefits, and studies.

Second, access to annual fee collections could support activities such as the development of interim storage facilities, establishment of the transportation system, siting and characterization of a geologic repository, and execution of regulatory development and oversight. This access could be accomplished either through legislative reclassification of fee collections from mandatory to discretionary, or as a direct mandatory appropriation of the fees, or some combination thereof. Legislative reclassification of fee collections from mandatory to discretionary would allow the fees to offset NWF discretionary appropriations, so that appropriation of the fees no longer would have to compete with other discretionary priorities. Instead, fees would be provided in amounts needed only above the annual appropriations described above and would also be limited by the amount of fee income, as envisioned by the NWPA. This approach could be preferable if additional Appropriator involvement was desired or deemed necessary and regular annual appropriations of that magnitude could be identified.

Alternatively, a direct mandatory appropriation of the annual fees could be coupled with direct access to the corpus of the NWF, as further discussed below. Under this arrangement, spending could be controlled through annual mandatory spending caps set by Congress or by tying funding levels to specific system development milestones in legislation. With continued oversight by the Appropriations Committees, these mandatory spending caps could be adjusted, as deemed necessary and appropriate. Implementation of either or a combination of both of these approaches will require substantial consultation with Authorizing, Budget, and Appropriations Committees of Congress; the Administration is committed to working with Congress to find a mutually agreeable solution to this issue.

Third, regardless of how access to the annual fees is provided, the substantial corpus of the NWF will be needed at an appropriate time in the future, particularly to support the development of a geologic repository. The cost of constructing repository facilities could outstrip the annual fee collections and other discretionary appropriations discussed above. Direct access to the corpus of the NWF through mandatory appropriations could be carefully managed by limiting its use to specific capital expenditures, tied to performance triggers, such as meeting licensing actions and major construction milestones, or subject to hard spending caps.

The cost of the government's growing liability for partial breach of contracts with nuclear utilities is paid from the Judgment Fund of the U.S. Government. While payments are extensively reviewed by DOE, and must be authorized by the Attorney General prior to disbursement by the Department of the Treasury, as mandatory spending they are not subject to Office of Management and Budget or Congressional approval. Past payments are included in full in the budget, but the budget does not reflect full estimates of the future cost of these liabilities and does not fully reflect the potential future cost of continued insufficient action. Future budget projections would be improved by including the full cost of estimated liability payments in the baselines constructed by both CBO and OMB. If the full cost of the estimated liability payments is accurately reflected in the baseline program costs over the life of the project would eventually be offset by reductions in liabilities as the government begins to pick up sufficient waste from commercial sites. As a result, the projected long-term cost of insufficient action surpasses the cost of implementing the program in the short run.

Any new funding structure for this program will need to balance increased funding flexibility and rigorous spending oversight to help assure that the program is implemented in the most cost-effective manner possible, while still holding the MDO accountable to the President and Congress. Further, crafting the MDO funding structure will require a creative and nuanced approach to providing needed funds with involvement by the Administration and all of the appropriate committees of Congress, working together to achieve a viable solution within the current federal budget rules and procedures.

The President's fiscal year 2014 budget will include additional details regarding funding for the program of work described in this Strategy document.

ONGOING ACTIVITIES

Within DOE, the Office of Nuclear Energy's Office of Fuel Cycle Technology has initiated a planning project with the objective of pursuing activities that can be conducted within the constraints of the NWPA and will facilitate the development of an interim storage facility, of a geologic repository, and of the supporting transportation infrastructure. The activities being conducted can be transferred to a new MDO when established and will not constrain its options. This includes initiating planning for a large-scale transportation program; evaluating operational options for consolidated storage and furthering the design of a generic consolidated storage facility. The Department is also developing plans for initiating a consent-based siting process. The Department will continue with these activities and those listed below, within existing Congressional authorization, while the Administration and Congress work together on potential changes to the nuclear waste management program.

The BRC also urged the Department to evaluate options for transportation of used nuclear fuel from shut down reactors. In 2013, DOE is evaluating the inventory, transportation interface, and shipping status of used nuclear fuel at shut-down reactor sites. The Department has established cooperative agreements with state and regional groups and engaged tribal representatives to begin discussions on transportation planning and emergency response training consistent with NWPA Section 180(c). Further, the Department is considering how best to leverage the work of state and regional groups currently engaged in transportation planning and oversight of radioactive waste shipments to WIPP in New Mexico.

In FY 2013, the Department is undertaking disposal-related research and development work in the following areas: an evaluation of whether direct disposal of existing storage containers used at utility sites can be accomplished in various geologic media; an evaluation of various types and design features of back-filled engineered barriers systems and materials; evaluating geologic media for their impacts on waste isolation; evaluating thermal management options for various geologic media; establishing cooperative agreements with international programs; and developing a research and development plan for deep borehole disposal, consistent with BRC recommendations.

CONCLUSION

In this Strategy, the Administration has highlighted agreement with many of the principles of the BRC recommendations and has outlined actions that, with legislative authorization by Congress, can lead to a safe and responsible solution to managing the nation's nuclear waste. Indeed, action by Congress in the form of new authorizing legislation and appropriations is necessary for success of the waste management mission. Specifically, legislation is needed in the near term to permit or address the following activities over the next 10 years:

- Active engagement in a broad, national, consent-based process to site pilot and full-scale interim storage facilities, and site and characterize a geologic repository;
- Siting, design, licensing, and commencement of operations at a pilot-scale storage facility with an initial focus on accepting used nuclear fuel from shut-down reactor sites.;
- Significant progress on siting and licensing of a larger consolidated interim storage facility capable of providing system flexibility and an opportunity for more substantial progress in reducing government liabilities;
- Development of transportation capabilities (personnel, processes, equipment) to begin movement of fuel from shut-down reactors;
- Reformation of the funding approach in ways that preserve the necessary role for ongoing discretionary appropriations and also provide additional funds as necessary, whether from reclassified fees or from mandatory appropriation from the NWF or both; and
- Establishment of a new organization to run the program, the structure and positioning of which balance greater autonomy with the need for continued Executive and Legislative branch oversight.

This Strategy translates the BRC's report and recommendations into a set of broad steps that will ultimately benefit the entire nation. The Administration will work closely with Congress to develop a path forward that maximizes the likelihood of success. When executed, the new program will provide near-term and long-term solutions for managing the back-end of the nuclear fuel cycle, thereby resolving a longtime source of conflict in nuclear policy by providing safe, secure, and permanent disposal. Until the necessary new legislation has been enacted, the Administration will pursue components of the Strategy as described above pursuant to current law and in close coordination with Congress. Finally, in executing the program the federal government must work closely with potential host states, tribes, and communities whose engagement will be essential for successfully operating a comprehensive used nuclear fuel and high-level radioactive waste storage, transportation, and disposal system.

End Notes

[1] The term "used nuclear fuel" as used in the BRC charter and in this document is intended to be synonymous with the term "spent nuclear fuel" as used in the Nuclear Waste Policy Act and the Standard Contracts.

[2] The BRC recommendations are summarized as follows:

1. A new, consent-based approach to siting future nuclear waste management facilities.
2. A new organization dedicated solely to implementing the waste management program and empowered with the authority and resources to succeed.
3. Access to the funds nuclear utility ratepayers are providing for the purpose of nuclear waste management.
4. Prompt efforts to develop one or more geologic disposal facilities.
5. Prompt efforts to develop one or more consolidated storage facilities.
6. Prompt efforts to prepare for the eventual large-scale transport of used nuclear fuel and high-level waste to consolidated storage and disposal facilities when such facilities become available.
7. Support for continued U.S. innovation in nuclear energy technology and for workforce development.
8. Active U.S. leadership in international efforts to address safety, waste management, non-proliferation, and security concerns.

[3] J. C. WAGNER et al., Categorization of Used Nuclear Fuel Inventory in Support of a Comprehensive National Nuclear Fuel Cycle Strategy, ORNL/TM-2012/308 (FCRD-FCT-2012-00232), Oak Ridge National Laboratory, Oak Ridge, Tenn., December 2012.

[4] Choosing a New Organization for Management and Disposition of Commercial and Defense High-Level Radioactive Materials, RAND Corporation, Washington, DC, MG-1230-DOE, 2012. The report is available free for downloading at www.rand.org/pubs/monographs/MG1230.html.

INDEX

D

Q

R

S

T

U

V

W

Y